S0-HIO-677

Brolin, Wettermark

Bioluminescence Analysis

VCH

© VCH Verlagsgesellschaft mbH, D-6940 Weinheim (Federal Republic of Germany), 1992

Distribution:
VCH, P.O. Box 101161, D-6940 Weinheim (Federal Republic of Germany)
Switzerland: VCH, P.O. Box, CH-4020 Basel (Switzerland)
United Kingdom and Ireland: VCH (UK) Ltd., 8 Wellington Court, Cambridge CB1 1HZ (England)
USA und Canada: VCH, Suite 909, 220 East 23rd Street, New York, NY 10010-4606 (USA)

ISBN 3-527-28194-0 (VCH, Weinheim) ISBN 0-89573-995-X (VCH, New York)

S. Brolin, G. Wettermark

Bioluminescence Analysis

VCH Weinheim · New York · Basel · Cambridge

574.19125
B787B

DEC 17 1992

Prof. Dr. Sven Brolin
Uppsala University
Dept. of Medical Cell Biology
Box 571
S-75123 Uppsala
Sweden
Phone: -46-18-174000
Fax: -46-18-556401

Prof. Dr. Gunnar Wettermark
Head of Research
COR AB
Ekbacksvägen 1
S-18233 Danderyd
Sweden
Phone: -46-8-7550729

This book was carefully produced. Nevertheless, authors, editors and publisher do not warrant the information contained therein to be free of errors. Readers are advised to keep in mind that statements, data, illustrations, procedural details or other items may inadvertently be inaccurate.

1st edition 1992

Published jointly by
VCH Verlagsgesellschaft mbH, Weinheim (Federal Republic of Germany)
VCH Publishers, Inc., New York, NY (USA)

Editorial Director: Dr. Hans F. Ebel
Editorial Manager: N. Banerjea-Schultz
Production Manager: Dipl.-Wirt.-Ing. (FH) Bernd Riedel

Library of Congress Card No.: 91-30270

British Library Cataloguing-in-Publication Data
Brolin, Sven
Bioluminescence analysis.
I. Title II. Wettermark, Gunnar
574.19
ISBN 3-527-28194-0

Die Deutsche Bibliothek – CIP-Einheitsaufnahme
Brolin, Sven:
Bioluminescence analysis / S. Brolin ; G. Wettermark. – 1. ed.
– Weinheim ; New York ; Basel ; Cambridge : VCH, 1992
ISBN 3-527-28194-0 (Weinheim…)
ISBN 0-89573-995-X (New York)
NE: Wettermark, Gunnar:

© VCH Verlagsgesellschaft mbH, D-6940 Weinheim (Federal Republic of Germany), 1992

Printed on acid-free and low-chlorine paper.

All rights reserved (including those of translation into other languages). No part of this book may be reproduced in any form – by photoprinting, microfilm, or any other means – nor transmitted or translated into a machine language without written permission from the publishers. Registered names, trademarks, etc. used in this book, even when not specifically marked as such, are not to be considered unprotected by law.

Composition: Mitterweger Werksatz GmbH, D-6831 Plankstadt
Printing: betz-druck gmbh, D-6100 Darmstadt 12
Bookbinding: J. Schäffer, D-6718 Grünstadt

Printed in the Federal Republic of Germany

Preface

Today, small quantities of light can be measured accurately at low cost. This is one of the reasons why fluorescence and phosphorescence have become widely accepted analytical tools. Similarly, the use of chemically produced light (chemiluminescence) is rapidly progressing as it offers the additional advantages of being independent of external excitation. On the other hand, the chemical production of light requires a higher level of attention to the reaction conditions. In bioluminescence analyses, the molecules to be detected play a key role in the reactions leading to the production of light. Thus, ATP may be analyzed using the firefly system, as it reacts with firefly luciferin in the path towards excited oxyluciferin, which emits light. Analytical utility has been ascribed to many light-producing systems, although some of these are not suited to immediate practical application. It is, furthermore, a common experience that side reactions occur, which also give rise to light emission. This may impede high sensitivity applications.

Methods suitable for biochemical analyses of small objects are useful in many fields of biological and biochemical research. Studies at high sensitivity are particularly desirable if the cells to be analyzed are available only in small quantities as when samples from fine needle or punch biopsies have to be divided between routine diagnostics and biochemical studies. The bioluminescence technique is eligible for this purpose due to a high quantum yield, specificity of the involved enzymes, and the possibilities of linking a great variety of metabolic reactions to a final bioluminescent step. In the analytical performance attention has to be drawn to the appropriate formation and consumption of the nucleotide form which frequently serves as the product to be determined. The results of ATP determinations, for instance, are influenced by the state of equilibrium in the adenylate pool. Another example is the uncontrolled consumption or extraneous formation of NADH during coupling to selected dehydrogenase reactions.

This book is an attempt to outline the analytically relevant chemical processes behind the formation of cold light and to elucidate the special kinetic conditions which have to be taken into account for the design of bioluminescence analyses of nucleotides, enzymes, and substrates. The presentation is, furthermore, focused on practical aspects, such as how to prepare specimens and reagents in order to make full use of the inherent advantages of the bioluminescence methods, and how to avoid many pitfalls, which have sometimes hampered the analytical performance.

Sven E. Brolin
Gunnar Wettermark
July 1991

Table of Contents

1 The Phenomenon of Chemiluminescence

"Historically, studies of bioluminescence were of a scholarly nature directed toward an understanding of mechanisms involved in the operation of excited states of molecules. During the past years, however, it has become apparent that bio- and chemiluminescence are powerful tools for assaying a variety of biologically important molecules" [1].

1.1 The Nature of Chemiluminescence

1.1.1 The Concept

Light emission from electronically excited species is known as luminescence. The mode of excitatation of the energy-rich species is usually indicated by means of a prefix, e.g. triboluminescence (crushing of crystals), pyroluminescence (heat), electroluminescence (electricity). Chemiluminescence means that excitation is produced in the course of chemical reactions. It is a matter of energy transfer during the chemical reaction leading to the emission of light quanta. Luminescence is often described as the production of "cold light", as the chemical energy is converted into light without the intermediate production of heat. However, heat may be produced in the overall reaction simultaneously with light production. The phenomenon of chemiluminescence accompanies many kinds of reactions, most of them having an oxidative step in common. Some living organisms are capable of producing light and this phenomenon is denoted bioluminescence. It is thus a subgroup of the general phenomenon of chemiluminescence. However, in some literature the word chemiluminescence is given a more limited meaning, denoting only non-biological, chemical production of light.

Disregarding the complexity of reactions behind the production of a photon emitter, the light-producing process may by presented as an activation step, leading to production of an electronically excited compound, followed by a degradation giving rise to photons (Figure 1-1). Realizing that the scheme of Figure 1-1 only summarizes the

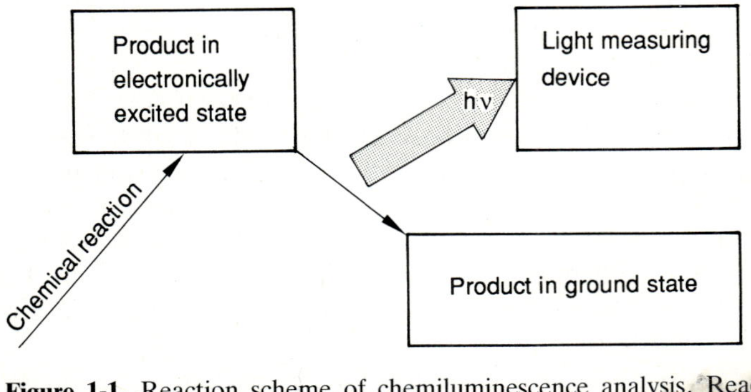

Figure 1-1. Reaction scheme of chemiluminescence analysis. Reactants which give rise to a product in an electronically excited state can be determined, if the return to the ground state is associated with the emission of light. The analysis applies to a great variety of molecules, including synthetic compounds and species obtained from biological extracts [2].

events in the production of light, it may serve as a basis for the understanding and evaluation of current luminescence analyses. However, a more detailed description of the chemiluminescence process will be presented in the next sections, in order to meet wishes from readers, who are interested in analytical refinements. Our knowledge is largely derived from experiments with synthesized nonbiologically organic compounds. Studies of the reaction mechanisms in bioluminescence started much later, when the analytical applicability of chemiluminescence had been well established and luminescent biological compounds had been isolated and identified.

1.1.2 The Excitation

Detailed understanding of the excitation mechanism in chemiluminescence is limited to a few reactions occurring in the gas phase. However, in general terms all chemical reactions may be imagined as occurring along a reaction coordinate as depicted in Figure 1-2 for the reaction:

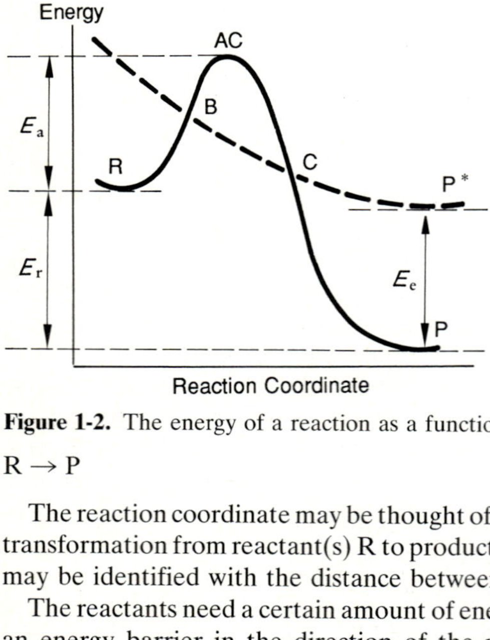

Figure 1-2. The energy of a reaction as a function of the reaction coordinate.

$$R \rightarrow P \quad \text{(R 1-1)}$$

The reaction coordinate may be thought of as a qualitative description of the extent of transformation from reactant(s) R to product(s) P. In a dissociating diatomic molecule it may be identified with the distance between the two atoms.

The reactants need a certain amount of energy, the activation energy E_a, to overcome an energy barrier in the direction of the reaction coordinate. This barrier may be thought of as a high energy transition-state or activated-complex AC, which may break down in the direction of the coordinate to form P. The rate of the reaction is proportional to the concentration of AC, which in an ordinary thermal reaction is set by the temperature following the Boltzmann distribution. Such a reaction may be considered as taking place in the same electronic state, represented by the solid line in Figure 1-2. It is usually the singlet ground state, S.

If the amount of the energy, E_r, which may be released in the reaction, is of suffient magnitude it may be sufficient to form a product in an electronically excited state. In the orbital model one electron is lifted to a more energy-rich orbit during the course of the reaction. Electronic excitation as part of a chemical reaction is an unlikely reaction path for most chemical systems even if the latent energy is sufficient. However, such a process may occur when conditions are set right. One prerequisite is that there exists a reaction path for an electronically excited state which crosses that of the ground state. In the cross-over point the molecular conformation is the same, providing for a possibility for change of state. This is depicted in Figure 1-2 with the dotted line indicating an excited singlet state. There is thus an alternative reaction path

$$R \rightarrow B \rightarrow C \rightarrow P^* \quad \text{(R 1-2)}$$

resulting in a product in a singlet excited state P*, which competes with the ordinary thermal reaction path

$$R \rightarrow AC \rightarrow P. \quad \text{(R 1-3)}$$

P* loses its excitation energy forming the product in the ground state P. The energy may be released as heat, in a non-irradiative process, but may in some cases be emitted as a chemiluminescence photon. Thus a basic scheme for luminescence may be written as:

$$R \rightarrow P^* \quad \text{(R 1-4)}$$

$$P^* \rightarrow P + h\nu \quad \text{(R 1-5)}$$

The energy difference between P* and P determines the frequency ν, and hence the color, of the emitted light:

$$E_e = h\nu \quad (1\text{-}1)$$

1.1.3 Emission and Quenching

The light resulting from chemiluminescence (or bioluminescence), is of the same nature as fluorescence and phosphorescence. In the model used in Figure 1-2 with a singlet ground state and an excited singlet state, the chemiluminescence is identical to the fluorescence of P, which means that the same kind of emission is possible to bring about by directly activating P to P* with an appropriate radiation source. P* may be deactivated in many other ways than by emitting a photon, i.e. in the fluorescence process. It may lose its electronic energy by a radiationless transformation to P or react chemically. The radiationless transitions are favored by collisions and other interactions with neighboring molecules. Usually this results in a lowering of the emission ("quenched emission"). Molecules acting in this way are called quenchers and the remaining emission is determined by the relative efficiency of the fluorescence path in relation to the dark processes, e.g. in the case of a quencher Q, reaction (R 1-5) in relation to reaction (R 1-6):

$$P^* + Q \rightarrow P \quad \text{(R 1-6)}$$

Many molecules are extremely efficient quenchers and may totally extinguish the emission. However, there are also molecules which may absorb the energy and become electronically excited and capable of emitting light. In fact with some chemiluminescent systems direct luminescence is not possible as P* is rapidly inactivated in dark processes.

In these cases only indirect luminescence is possible, letting the energy being rapidly transformed to a suitable emitter E.

$$P^* + E \rightarrow P + E^* \qquad \text{(R 1-7)}$$

$$E^* \rightarrow E + h\nu \qquad \text{(R 1-8)}$$

By selecting an emitter other than P*, the spectral distribution and light intensity may be changed. It is always a matter of competing processes, where high light production parallels efficient energy transfer and emission yield in the direction of the light producing path – in relation to all other paths.

1.1.4 The Photon Emitter

In some natural systems, i.e. in bioluminescence, the excited molecule, which emits the photon, may be identical with the product P of the exergonic chemical reaction. This occurs in the firely, which thus represents a case of direct chemiluminescence. In other organisms the chemically excited molecule may transmit its energy to a special emitter molecule. *Renilla reniformis* is an example of such indirect chemiluminescence, cf. Table 9-1. Sometimes the expression sensitized chemiluminesce is used as a synonym for indirect luminescence and the word sensitizer used for the emitter. This wording stresses the importance of the emitter molecule in its functions of absorbing the available energy of the excited species and transforming it into light. In vitro it is sometimes possible to enhance the emission of a bioluminescent system by adding a compound with efficient fluorescence properties. The excitation energy is transmitted to this compound which now emits with its characteristic spectrum instead of the natural emitter. This procedure is occasionally called the adding of "stain" to the reaction composition.

For a more detailed understanding of the emission process it is suitable to use a Jablonski diagram, which may apply both to the product P or an emitter E. Using this type of diagram the possible transitions between the ground state and the various excited states are illustrated in Figure 1-3. The energy is given on the ordinate and the

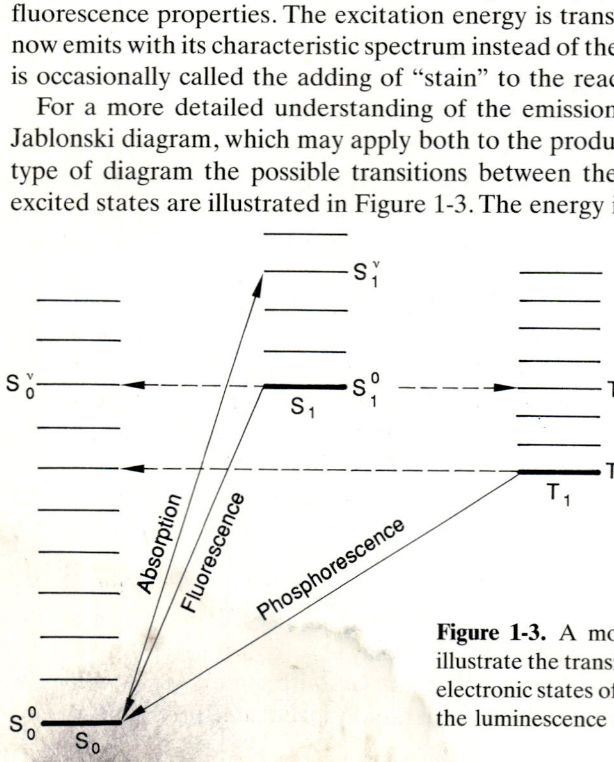

Figure 1-3. A modified Jablonski diagram, used to illustrate the transitions between the energy levels and electronic states of the product, P, or the emitter, E, of the luminescence reaction.

electronic states are indicated together with their vibrational levels. The diagram shows the situation for a molecule with a singlet ground state S_0, and a singlet excited state S_1. The molecule is also considered to have a triplet state T_1, where the triplet T_1 has a lower electronic energy than than S_1. Possible radiative processes between the states are indicated by straight arrows and non-irradiative (dark) processes with dashed arrows. Dark transitions between states of the same multiplicity ("same spin") may occur, i.e., internal conversion, and also between states of different multiplicity, i.e., intersystem crossing. Generally, transitions between states of different multiplicity have low probabilities and represent a case of "forbidden" transition, whereas transitions between states of the same multiplicity are "allowed", from a spin point of view. However, the recognition of forbidden and allowed transitions is not always distinct and many forbidden transitions can have a high probability. The term fluorescence is usually used to indicate a light-emitting process as part of an allowed transition, whereas phosphorescence is thought to emanate from a forbidden transition. Thus in connection with change of spin, fluorescence is commonly defined as a light-emitting transition between states of the same multiplicity and phophorescence between states of different multiplicity. Following excitation, fluorescence, being an allowed transition, is generally of short duration, whereas the phosphorescence is more long-lived, reflecting the life times of S_1 and T_1, respectively.

Let us now use the Jablonski diagram (Figure 1-3) to illustrate some properties of a product P of the chemiluminescence reaction, as given in Figure 1-2. The chemical reaction was considered to produce P*, i.e. P in the singlet excited state S_1. It is likely to be produced with some vibrational (and rotational) activation S_1^v, which, at least in the condensed phase, is rapidly lost to give S_1^0. Obviously the same state S_1^0, may be produced by exciting S_0 directly, using radiation which is absorbed by S_0, yielding S_1^v, with the vibrational energy dependent on the wavelength of the exciting light. There are now three possible means for deactivation of S_1^0: fluorescence, internal conversion to S_0^v, or intersystem crossing to T_1^v. If fluorescence occurs this process will yield the light emitted at chemiluminescence. Despite being spin-forbidden the intersystem crossing may be a very efficient process competing favorably with the other reaction paths. Once formed the vibrationally excited triplet T_1^v degrades rapidly to T_1^0, and chemiluminescence, emanating from phosphorescence, is observed, if the radiative process from T_1 to S_0 is not obscured by other more efficient deactivation processes.

In principle, it should be possible to include all the various electronic states in Figure 1-2 and draw the energy curve along the reaction coordinate. In the above example it was assumed that the product P was formed in the singlet excited state; i.e. that $P^* = S_1$, defining the position of entry into the Jablonsky diagram in Figure 1-3. There may be systems with a singlet ground state, where the chemical reaction directly results in a triplet state and this will obviously have consequences on the further reaction paths. In case of energy transformation to a particular emitter, reaction (R 1-7), it is neccessary that the emitter possesses a suitable electronic state with a lower energy than the primary excited species. In the Jablonski diagram this means that P should have an S_1-level above that of E, if $P^* = S_1$, but the Jablonski diagrams may otherwise be completely different.

1.1.5 Quantum Yield

The effiency of the conversion of chemical energy into light is given by the quantum yield (= quantum efficiency). The overall quantum yield, Φ_{CL}, of chemiluminescence is defined as the ratio of the number of photons emitted and the number of molecules participating in the chemiluminescence reaction. It is thus a product of the efficiency, Φ_{CR}, of the chemical reaction leading to the light emitting excited state, multiplied by the efficiency of light emission from this state, Φ_{FL}.

$$\Phi_{CL} = \Phi_{CR} \cdot \Phi_{FL} \qquad (1\text{-}2)$$

In the example given above, where a fluorescent singlet is produced, it is the effiency of production of the singlet, S_1, multiplied by the quantum yield of fluorescence. (Note that absorption of light by S_0 means formation of S_1 with a quantum yield of unity.) It is thus possible to obtain information about the efficiency of the chemical excitation step by measuring the overall efficiency of a chemiluminescence process and the quantum yield of fluorescence. This information can be used to assess the potential of sensitizing with a more efficient fluorescer than the original emitter. However, fluorescence quantum yields, Φ_{FL}, are often high, in particular for natural systems, i.e. in bioluminescence, and the limitations towards obtaining a high quantum yield often lie in the chemical step.

It should be realized that the formation of the emitter, even in the case of a non-sensitized reaction, involves several reaction steps where highly reactive species are involved. This means that there are ample provisions for side-reactions of various kinds and deactivation during the path towards production of the emitter. Also the emitter may be subject to efficient deactivation in non-radiative processes, as experienced in studies of fluorescence and phosphorescence. In particular, in solution there are a great many possibilities for deactivation of excited molecules, due to the randomized nature of the processes with freely dispersed and mobile molecules. Therefore, quantum yields of chemiluminescence are generally low and there are only few systems known, where the conditions are such that the loss factors can be minimized and that the steps leading to light productions become the predominating path. This is notably the case with decompositions of oxalyl esters which in the presence of a sensitizer can be made efficient enough to be used for illumination purposes, cf. Section 1.2.

In bioluminescence the excited molecules may be rather restricted with respect to their reaction pathways due to binding at enzymatic sites. Nature has in this way developed very efficient luminescence reactions with high quantum yields.

1.2 Chemiluminescent Reactions

1.2.1 Discovery of Luminescent Compounds

For a long time biological systems, being the first chemiluminescent systems to be studied, did not provide enough material for chemical characterizations. However, with the discovery of light-producing organic reactions from well-defined compounds, reaction mechanism studies could be initiated. As early as 1877 Radziszewski [3] observed that lophine (2,4,5-triphenylimidazole) emitted green light when reacting with oxygen in alkaline solution and that benzaldehyde was formed during the reaction. This discovery invited further identification of final reaction products and this type of activity formed an essential part of the early research in the field. Equipment soon became available to make an evaluation of quantum yields possible. Questions about short-lived intermediates arose at a much later date. The reaction mechanism operating in this type of compounds is nowadays fairly well understood, cf. Figure 1-4. It is a

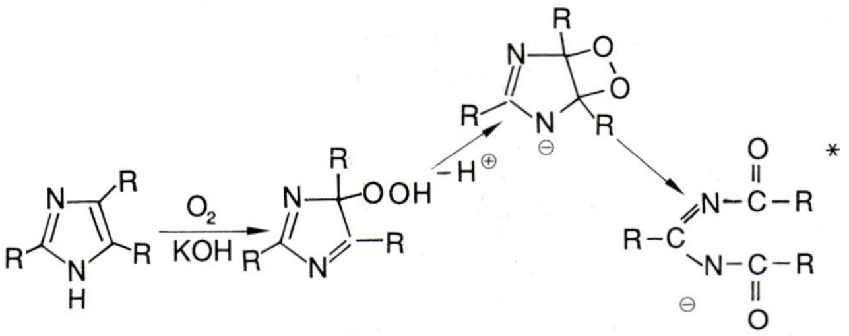

Figure 1-4. The mechanism operating in the lophine-type compounds.

matter of base-catalyzed oxidation of an imine to a corresponding dibenzoylbenzamide and it is the excited singlet of the benzamide anion that emits the light. The first step in the oxidation is the formation of a hydroperoxide and it appears that the further process proceeds via a short-lived dioxetan.

1.2.2 Luminol

In particular, studies on the compound luminol (5-amino-2,3-dihydro-1,4-phthalazinedione) have played an important role in the understanding of the processes underlying the chemical induction of light emission. The light-producing oxidation was originally described by Albrecht in 1928 [4]. This light conversion proceeds with remarkably high efficiency, which facilitated and stimulated further studies. Under favorable conditions the quantum yield of luminescence may be $\Phi_{CL} = 0.015$ with $\Phi_{CR} = 0.05$ and $\Phi_{FL} = 0.30$ [5].

The attention paid originally to luminol and its reactions was strengthened by various spectacular effects, which attracted much interest. One was an analytical application, in which the catalytic action of hemin was used to detect minute amounts of blood. Specifically, in forensic chemistry specks of blood on a piece of cloth could be revealed, even if it had been carefully washed, by spraying it with an alkaline solution of luminol and hydrogen peroxide.

Luminol represents a class of compounds which opened up new possibilities of using synthetic means for studying effects at the reaction site. Several steps of the luminescent reaction are known, and have guided studies of other luminescent systems; but further details are continuously revealed [6]. The emitter has been identified as the 3-aminophthalate ion. The reaction occurs both in water and in aprotic solvents (e.g. dimethyl sulfoxide), but the detailed reaction mechanisms are different. In water a catalyst is needed and this forms the basis for a variety of analytical applications. Suggested intermediates include a hydroperoxide and an endoperoxide (Figure 1-5).

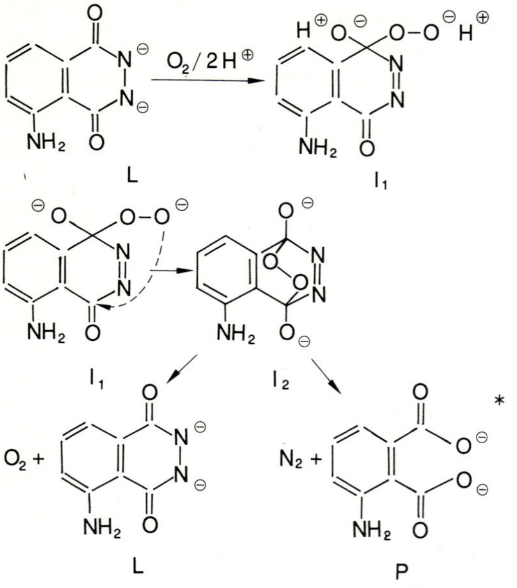

Figure 1-5. Basic outline of a postulated mechanism for the luminescence reaction during the oxidation of luminol. In the overall reaction, (luminol = 5-amino-2,3-dihydro-1,4-phthalazinedione, (L)), by cleaving molecular nitrogen, enters an excited state of 3-aminophthalate (P), which is the light emitter. The primary intermediate of the oxidation is an open hydroperoxide (I_1). When it is deprotonated, the hydroperoxide group reacts in a nucleophilic attack on the carbon of the carbonyl group to form a bicyclic endoperoxide (I_2). The endoperoxide decomposes spontaneously, liberating oxygen or nitrogen. When oxygen is expelled, luminol reappears. This is the major pathway and is a dark reaction. When nitrogen is released (about 4 %), excited 3-aminophthalate is formed.

1.2.3 A Wide Variety of Chemiluminescent Systems

The research on luminescence is broad and many systems are being investigated [7-10], some of which will probably be of analytical importance. In fact many systems are already being used in routine analyses, in particular with respect to pollution of air. It should be emphasized that several of the non-biological systems also find analytical application in biochemistry. Their use as markers in various contexts may be particularly interesting, e.g., for labelling antibodies and antigens in luminescence immunoassay. Derivatives of acridine are examples of luminescent compounds well-suited for such purposes, because of their good luminescence efficiency under relatively mild conditions, requiring only hydrogen peroxide and dilute alkali [11, 12]. The quantum yield Φ_{CL} may approach 5 % and the mechanism of light-production is fairly well understood, cf. Figure 1-6. The emitting excited product is *N*-methylacridone. A molecule containing a dioxetanone ring appears to be the crucial intermediate. Excited *N*-methylacridone is also formed in the chemiluminescence reaction of lucigenin (bis-*N*-methylacridinium nitrate). This molecule is probably, together with luminol, the most studied chemiluminescent compound. Again there are strong indications that the reaction proceeds via a cyclic peroxide, in this case dioxetan, which, being very

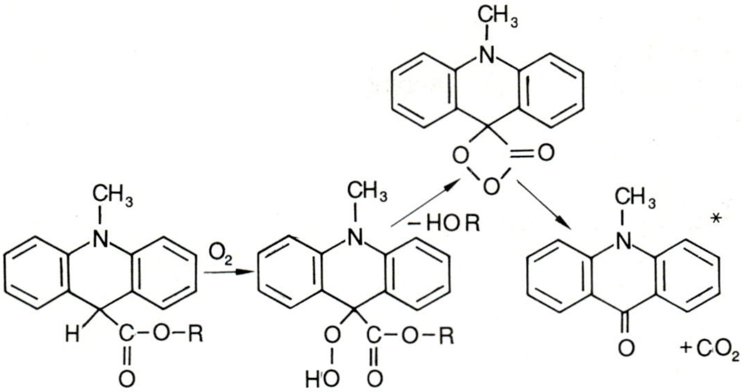

Figure 1-6. Basic scheme for luminescence of acridan

energyrich, decomposes spontaneously yielding the electronically excited carbonyl compound (*N*-methylacridone).

The hitherto most efficient chemiluminescent systems, based on synthetic compounds, have been obtained with certain oxalates. A great variety of such systems are known with $\Phi_{CL}>0.2$. They are primarily the result of an effort to obtain "chemical light" to be used for emergency lighting and similar purposes, and are commercially available as "light rods". A light rod is a small transparent cylindrical container with two-compartments separated by a seal which can be broken to let the reactants mix and yield a chemiluminescence reaction. One compartment may suitably contain the oxalate and a fluorescent dye, e.g. rubrene, and the other hydrogen peroxide and an organic base, cf. Figure 1-7. The reaction is an oxidation of the oxalate with hydrogen

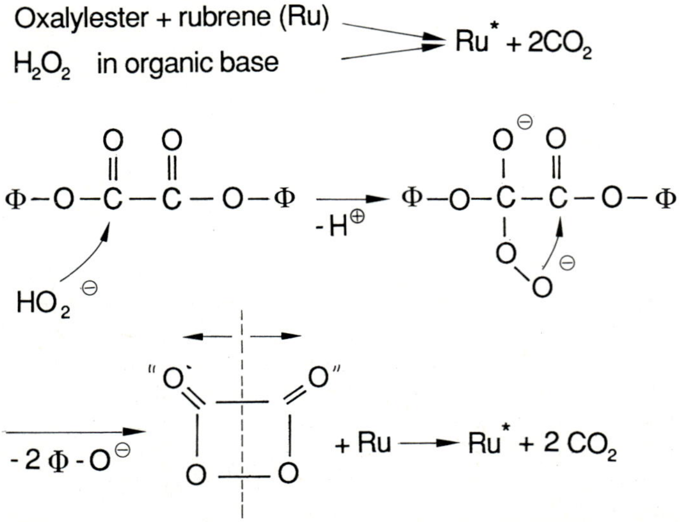

Figure 1-7. An example of a sensitized chemiluminescent system in which the energy from the oxidation of an oxalyl ester is transferred to an efficiently fluorescing hydrocarbon (rubrene). It is assumed that a "double" nucleophilic substitution is the initial step, followed by decomposition of the dioxetanone ring, and transfer of the energy to rubrene; probably through a radical-radical reaction.

peroxide in which the energy is transferred to the fluorescent dye. By varying the dye it is possible to obtain rods luminescing in different colors. Similar systems to those used in the light rods are also used for analytical purposes, e.g. in immunoassay and for the analysis of hydrogen peroxide [13].

1.2.4 Common Reaction Mechanisms

Reaction mechanisms of chemiluminescence often involve peroxide decompositions of various kinds. In the terminal reactions ion radicals may be involved, in a way similar to that described for electrochemiluminescence, cf. Section 1.2.5. In fact, chemiluminescence of organic compounds almost invariably involves oxidation with molecular oxygen or hydrogen peroxide. Sufficient energy becomes available for production of a visible light quantum when a weak O-O bond is replaced by a strong C=O bond. In order to produce a visible light quantum the available energy must exceed 190 kJ mole^{-1}, which is the energy of red light. Blue photons require as much as 310 kJ mole^{-1} assuming that the transition occurs without losses.

It was predicted from energetic and quantum mechanical considerations that dioxetan compounds should yield electronically excited states when thermally decom-

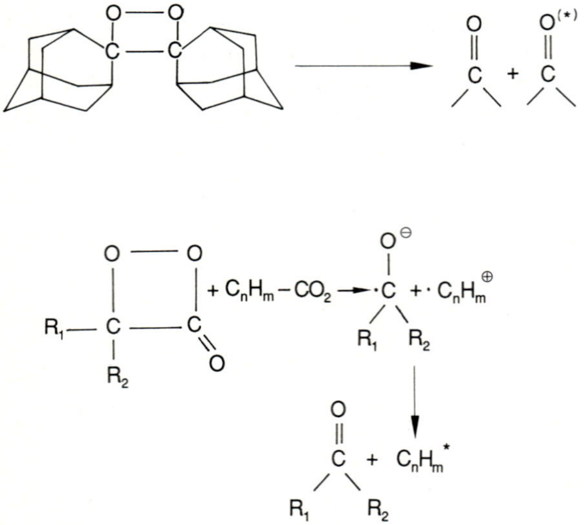

Figure 1-8. The dioxetan ring represents a molecular arrangement which is commonly seen in many bioluminescence reactions [20]. This peroxide ring is very energy-rich and splits spontaneously in such a way that the energy released can be transferred to an electronically excited state capable of emitting light. The picture also shows the structure of diadamantylendoperoxide, which is stable at room temperature. Compounds containing an oxetanone ring are even more unstable than the oxetans and decompose in a similar way.
The excited emitter may be the remains of the peroxide compound – as with the firefly – or another emitter molecule, already present or added. A sensitized decomposition may sometimes lead to a very efficient transfer of the energy to light. This is illustrated for a dioxetanone compound reacting with a hydrocarbon fluorescer. An initial formation of ion radicals is assumed, followed by a radical-radical annihilation step, in which the excited hydrocarbon is produced.

posing [14]. This was later confirmed and much of our current knowledge about chemical excitation comes from dioxetan chemistry. Stable compounds of this type may be synthesized using very large substituents to the dioxetan ring. Diadamantylendo-peroxide (adamantane-2,3′-[1,2] dioxetan-4′,2″-adamantane) is a good example (Figure 1-8). With simple dioxetans a major portion of the excited product is formed as a non-phosphorescing triplet. However, this is accompanied by the formation of a minor fraction of a fluorescing singlet. Furthermore, the singlet portion may be increased by suitably substituting the dioxetan, e.g. by providing it with easily oxidizable groups. Here the acridan derivative lucigenin serves as a good example (Figure 1-6). The yield of singlet also increases in the presence of a suitable catalyst like silica gel.

Compounds containing a dioxetanone ring are much more unstable than the corresponding dioxetan and are thus more difficult to isolate. A few containing very large substituents R_1 and R_2 have been synthesized and characterized using fast procedures. The dioxetanones have a strong disposition for liberating carbon dioxide, hence providing sufficient energy for leaving products in electronically excited states, i.e. in a similar way to that described for the dioxetans. The yield of excited singlet is basically poor but may be strongly increased by providing an aromatic hydrocarbon with low redox potential. In this way the dioxetanone ring, while decarboxylating, can accept an electron from the hydrocarbon and form ketyl and hydrocarbon radicals, cf. Figure 1-8. The two are formed close to each other and the "cage effect" makes them annihilate each other transforming the liberated energy into excitation of the hydrocarbon. It is thus the characteristic hydrocarbon fluorescence that is observed during these chemiluminescence reactions. The above mechanism is thought to operate during the reaction of an oxalate with hydrogen peroxide in the presence of a fluorescent dye, cf. Section 1.2.3.

1.2.5 Electrochemiluminescence

A great deal of information about chemiluminescence mechanisms has been obtained by producing reactive species by other means [6, 15]. Early on such experiments with luminol were conducted by Harvey 1927 [16]. It appears that the reactions leading to the production of light are the same as those involved in the purely chemical reactions.

Particularly strong luminescence is obtained when electrolyzing polycyclic hydrocarbons (A_r) in a suitable electrolyte. A radical-cation is formed at the anode and a radical-anion at the cathode. When the two ions combine, a light-emitting, electronically excited state is formed. Using such molecules, where the formation of a dimer is sterically hindered, the emitted spectrum is identical to that obtained in fluorescence. Others, which are not hindered, e.g., anthracene, also give rise to excimer luminescence, which means emission from the dimer, that is formed in the reaction. The electroluminescence reactions may thus be described as:

Anode $Ar \rightarrow \cdot Ar^-$ (R 1-9)

Cathode $Ar \rightarrow \cdot Ar^+$ (R 1-10)

Chemiluminescence $\cdot Ar^- + \cdot Ar^+ \rightarrow Ar^* + Ar$ (or $Ar_2{}^*$) (R 1-11)

1.3 Bioluminescence Mechanisms

1.3.1 Common Features

Bioluminescence has attracted the interests of authors and scientists throughout the centuries. Thus as early as in 1555 a book was published devoted wholly to this phenomenon [17]. A comprehensive historical review covering the time up to our century has been presented by Harvey [18] and is an established reference work.

In the last century, it was demonstrated that induction of bioluminescence required two components. One was named luciferase and was thought to exert an enzymatic action and assumed to induce the other compound, luciferin, to emit light [19]. The fascination of bioluminescence was enhanced by these early experiments and hence stimulated continued studies of light-producing reactions in living organisms. The early experiments concerned the click beetle *Pyrophorus* and the clam *Pholas* but it is now known that all systems cannot be described in such terms. However, with most systems studied the notations luciferase and luciferin apply and are used to denote the enzyme catalyzing the bioluminescent reaction and the oxidizable substrate. It is realized that the light emitter may be a completely different molecule than luciferin and even not a product of the reaction. It is thus a matter of direct or indirect chemiluminescence, cf. Section 1.1.3. The chemical structure of the the luciferin is known for several organisms. In some cases, notably bacteria, two substrates are oxidized (a long chain aldehyde and flavin mononucleotide) and may together be labelled as the luciferin of the reaction.

Particularly since the end of the second world war there has been an increasing interest in bioluminescence. Efforts were directed to clarifying the chemistry behind the biological photon emission and were stimulated by the progress in producing synthetic compounds having similar (and in some cases identical) structures to those involved in the bioluminescence reactions. Such compounds were made available in desired quantities, whereas the collection of sufficient amounts from luminescent organisms was hampered – for instance the preparation of 5 mg luciferin from *Renilla reniformis* required extraction from 30000 animals [20]. The overall work has resulted in detailed information about the chemistry and reaction mechanisms of the light emission of many species. A few example of this are given below.

1.3.2 Firefly

The reaction taking place in the firefly, *Photinus pyralis,* is a typical non-sensitized reaction, in which a luciferin is oxidized forming a fluorescent end-product, oxyluciferin (Figure 1-9). The excitation is extremely efficient, $\Phi_{CL} = 0.88$ [21]. The primary step is the reaction with ATP (in the form of a Mg-complex) to form luciferyladenylate. This reaction is catalyzed by luciferase. Once luciferyladenylate has been formed, all subsequent reactions in the light-producing process are non-enzymatic. The esterification of luciferin with ATP may be thought of as a means to provide a good "leaving group".

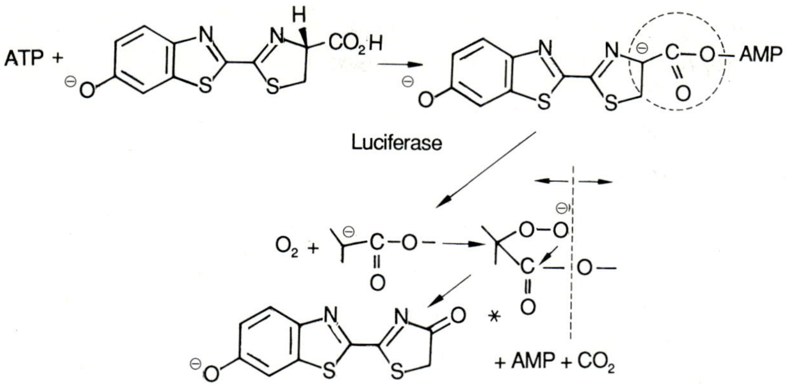

Figure 1-9. Some details in the mechanism of firefly luminescence. Luciferin is the molecule responsible for light formation.

The key step in the conversion of chemical energy into excitation energy takes place at the esterification site, which is marked by a dotted circle. Already at a neutral pH, luciferyladenylate exists in the anionic state; the hydrogen attached to the thiazol ring at the carbonyl substituent is acidic. Molecular oxygen reacts with this function to form a hydroperoxide. A dioxetan ring is built up by nucleophilic substitution, and adenylmonophosphate (AMP) is cleaved. The dioxetanone decomposes, expelling CO_2 and leaving excited oxyluciferin. Further details of the firefly reaction are given in Section 5.2, where it is viewed from the point of analysis.

1.3.3 Photobacteria

Marine luminous bacteria have attracted considerable research and yet several steps of their mechanism are unclear. There is a difference between the in vivo and in vitro luminescence both in terms of emission spectra and quantum yield [22]. It appears that it is a matter of sensitized luminescence with different emitters operating in the two cases. In vivo an efficient fluorescer, a lumazine derivative (cf. Figure 1-10) non-covalently bound to a low molecular weight protein, readily accepts the excitation energy, yielding $\Phi_{CL} \simeq 0.2$ *(Photobacterium leiognathi)*. As $\Phi_{FL} \simeq 0.5$ for the lumazine protein the chemical excitation must be very efficient, approaching $\Phi_{CR} \simeq 0.5$. The lumazine protein is not extracted along with the enzyme isolation resulting in less efficient sensitizers in the in vitro extract.

The substrates of the luciferase reaction include $FMNH_2$, a long chain aldehyde and oxygen. However, $FMNH_2$ is only transiently available in the bacteria and is produced by NADH acting on FMN. This preceding oxidoreductase reaction is often considered as part of the overall luminescence reaction and is particularly important in analytical applications. A presentation of the bacterial luciferase system from an analytical viewpoint is given in Section 5.3. The reaction mechanism is believed to involve a fast complexing of $FMNH_2$ with the luciferase and addition of oxygen to form an unstable hydroperoxide intermediate. In a subsequent step aldehyde enters and is oxidized. The liberated energy is transferred to a sensitizer. The pathway during this later part of the reaction is essentially unknown.

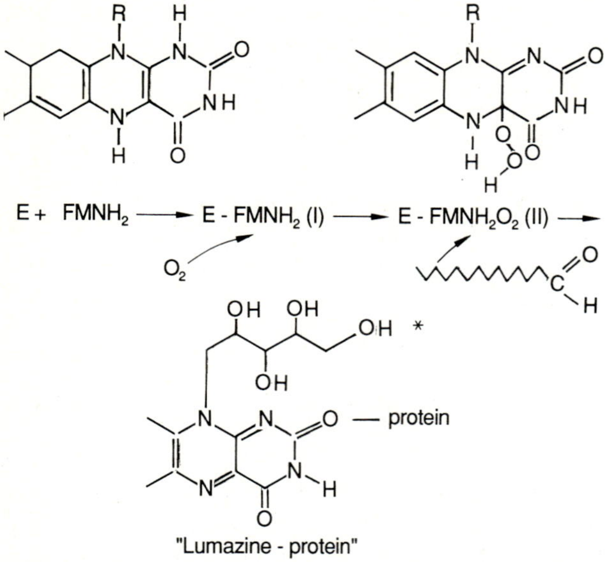

Figure 1-10. Some steps and molecules involved in the light-producing reactions of luminous bacteria. In vivo the emitter is a protein having the prosthetic group lumazine.

1.4 Analytical Applications

1.4.1 The Growth of Chemiluminescence Analysis

Looking back, some initial work was based on measurements of luminescence induced blackening of photographic emulsions. Although far from convenient, the method permitted quantitative recordings. Following the time course of the emission was easier when photoelectric techniques were introduced. The construction of sensitive photomultiplier tubes made it possible to monitor luminescence down to very low intensities. Electronic development and the availability of means for the immediate display of light production brought new facilities into luminescence research and prepared the ground for analytical developments.

Successful development and continuous usage of luminescence analysis in the gas phase have provided some basic insight into various problems connected with such analyses [23]. Luminescence is more common in the gas phase than in a condensed medium, because quenching reactions are less likely in the gaseous state. With atmospheric pollution becoming a matter of attention, it became urgent to develop sensitive methods for detecting minor quantities of substances; such as nitrogen oxides, sulfur, chlorine compounds, and ozone. Luminescence analysis in many cases proved adequate [8]. Further experience of analysis of airborne substances comes from the military sector where luminescence techniques were introduced to detect minute quantities of nerve gases [24].

During the last decades, rapid analytical development has also occurred regarding the use of luminescence in solution and in some cases also in the solid state [7, 8]. Here, access to newly synthesized chemiluminescent compounds has increased both the range of applicability and versatility [10]. In the fields of biology and medicine, the use of bioluminescence was largely preceded by analyses based on non-biological luminescent compounds. Thus, a favored technique for the detection of free oxygen radicals has been luminol-induced light emission. A widespread application is monitoring the "respiratory burst", i.e. the release of such radicals from granulocytes and monocytes during their immunological action [25, 26]. Another application is the use of a chemiluminescent compound as a label in immunochemistry. Examples of this are mentioned in Section 1. 2. 3. The chemiluminescent compound forms a ligand with the antigen or antibody and is developed by the addition of suitable reagents. Again, luminol is frequently used and the sensitivity may be increased, maybe up to a thousand times, by adding a sensitizer (cf. Section 1.1.3). However, luminol detection of oxygen radicals may also be used in immunological labelling techniques and provides a higher sensitivity than direct labelling with luminol. In such methods a suitable enzyme replaces luminol as a ligand, using an enzyme which catalyzes the production of oxygen radicals from the reagents. An example may be the identification of oligonucleotides in gene technology, where DNA probes are labelled with horseradish peroxidase [27]. Already without adding a sensitizer, the sensitivity is higher than that obtained with radioactive labelling, using ^{135}I.

1.4.2 Introduction of Bioluminescence

The description of cofactors and substrates for firefly luciferase by McElroy [28] opened a new approach to the assay of ATP (cf. Figure 1-9), drawing attention to the potentialities of bioluminescence. The new analytical possibilities were obviously of great interest to most biological sciences because ATP is the chief source of energy in all kinds of living cells. However, from the beginning, analytical application of firefly luciferase was severely hampered by ATP-production from contaminating ADP and adenylate kinase (cf. reaction (R 5-5). Adenylate kinase and ADP are difficult to remove from firefly extracts and their presence results in a disturbing background emission, due to long-lasting production of ATP. However, the analysis became useful and reliable, when the interfering production of ATP was prevented by degradation of ADP, through treatment with apyrase [29]. The problems involved in the interconversion of adenylates will be further treated in Section 5.2.2, particularly with regard to kinetic means for the detection and elimination of disturbances from contaminations in the reagents and samples.

It was early recognized that NADH delivered energy to the light production in luminescent bacteria [30]. The intermediate steps in the bacterial luciferase system were soon elucidated and also reviewed a few years later [31]. A couple of years later, its analytical applicability became apparent [32, 33]. The great possibilities of the bacterial luciferase system through coupling to dehydrogenase reactions was likewise demonstrated [33].

In addition to the firefly and the bacterial luciferase systems, there are a few others which have been of use for analytical purposes. The detection of Ca^{2+}, using activated photoproteins, in particular Aequorin, has been extensively reviewed by Blinks [34]. PAP (adenosine-3′-phosphate-5′-phosphate) may be determined using the luciferin-luciferase system of the sea-pansy, *Renilla reniformis* [35]. There are, however, a great many biological systems to which analytical applications have not been found but where such uses can be envisaged. A problem has sometimes been the limited availability of reagent material, which has, for instance, hampered the applications of Renilla preparations. With the development of gene technology and similar methods, this situation is likely to undergo dramatic changes [27, 36, 37].

Future development of bioluminescence analysis cannot simply be regarded as a continued extension of existing applications to various new areas. It may depend more on knowledge and progress in other fields, such as the processing of records from measurements of small quantities of light, the treatment of microdissected specimens or biological extracts, and acquaintance with the principles of enzymatic analysis. Thus, studies of bioluminescence alone will not permit the design of a satisfactory analytical procedure.

The following text is an attempt to present some relevant information needed for the successful application of bioluminescence analysis to microchemical studies in biochemistry, biology and medicine. The following chapters start with a brief description of the biochemical reactions and appropriate light production for analytical applications. Attention is paid to special kinetic conditions that are important for the analytical design. It is followed by discussions of the necessary steps involved in preparing a specimen for quantitative analysis, keeping the compound to be determined intact, and diminishing perturbing factors. The techniques for obtaining and recording the analytical signals are reviewed with reference to be procedure for obtaining high signal output and reproducible display even when light intensities are low. Finally, the diversity of the areas of application is illustrated, and examples are given of experimental models in which bioluminescence contributes to the solution of crucial problems.

References

[1] DeLuca MA. In: DeLuca MA, ed; Methods in Enzymology. New York: Academic Press; 1978 p. XIII (Bioluminescence and Chemiluminescence, Vol. LVII).

[2] Brolin SE, Wettermark G, Hammar H. Strahlenther. 1977; 153: 124–131.

[3] Radziszewski B. Ber Dtsch Chem Ges. 1877; 10: 70–75.

[4] Albrecht HO. Z Phys Chem. 1928; 136, 321–330.

[5] White EH, Roswell DF. In: Burr JG, ed; Chemi- and Bioluminescence. New York: Marcel Dekker; 1985: 215–244.

[6] Merényi G, Lind J, Eriksen TE. J Biolum Chemilum. 1990; 5: 53–56.

[7] Gundermann K-D. Chemilumineszenz organischer Verbindungen. Berlin: Springer Verlag; 1968: 1–174.

[8] Isacsson U, Wettermark G. Anal Chim Acta. 1974; 68: 339–362.

[9] Gundermann K-D. In: Schram E, Stanley P, eds. Analytical Applications of Bioluminescence and Chemiluminescence. Westlake Village Cal: State Printing & Publishing; 1979: 37–86.

[10] McCapra F, Perring KD. In: Burr JG, ed. Chemi- and Bioluminescence. New York: Marcel Dekker; 1985: 259–320.

[11] McCapra F, Watmore D, Patel A, et al. J Biolum Chemilum. 1989; 4: 51–58.

[12] Kinkel T, Lübbers H, Schmidt E, Molz P, Skrzipczyk HJ. J Biolum Chemilum. 1989; 4: 137–139.

[13] Arakawa H, Maeda M, Tsuji A. Chem Pharm Bull. 1982; 30: 3036–3039.
[14] McCapra F. Chem Commun. 1968: 155–156.
[15] Faulkner LR. In: DeLuca MA, ed; Methods in Enzymology. New York: Academic Press; 1978: 494–526. (Bioluminescence and Chemiluminescence, Vol. LVII).
[16] Harvey N. J Phys Chem. 1929; 33: 1456–1459.
[17] Gesner C. De raris et admirandis herbis quae sive quod noctu Juceant, sive alias ad causas, lunariae nominantur et obiter de alias etiam rebus, quae in tenebris lucent, commentariolus. Tiguri, 1555. Editione hac secunda emendatior, Hafniae, Copenhagen.
[18] Harvey EN. A History of Luminescence from the earliest times until 1900. Philadelphia: American Philosophical Society Independence Square; 1957.
[19] Dubois R. Compt Rend Soc Biol (Ser 8). 1885; 2: 559–562.
[20] Hori K, Cormier MJ. Biochim Biophys Acta. 1965; 102: 386–396.
[21] Seliger HH, McElroy WD. Arch Biochem Biophys. 1960; 88: 136–145.
[22] Lee J. In: Burr JG, ed. Chemi- and Bioluminescence. New York: Marcel Dekker; 1985: 401–437.
[23] Kenner RD, Ogryzlo ER. In: Burr JG, ed. Chemi- and Bioluminescence. New York: Marcel Dekker; 1985: 45–185.
[24] Goldenson J. Anal Chem. 1957; 26: 877–879.
[25] Allen RC, Stjernholm RL, Steele RH. Biochem Biophys Res Commun. 1972; 47: 679–684.
[26] VanDyke K, Castranova V. Cellular Chemiluminescence. Vols I, II, III. Boca Raton, Florida: CRC Press; 1987.
[27] Pollard-Knight DV. In: Stanley PE, Kricka LJ, eds. Bioluminescence & Chemiluminescence, Current Status. Chichester: John Wiley & Sons; 1991: 83–90.
[28] McElroy WD. Proc Natl Acad Sci USA. 1947; 33: 342–345.
[29] Lyman GE, De Vincenzo JP. Anal Biochem 1967; 21: 435–443.
[30] Strehler BL. J Am Chem Soc 1953; 75: 1264.
[31] McElroy WD, Seliger HH. Vol 25. In: Nord FF, ed. Advances in Enzymology. New York: Interscience Publ; 1963: 119–166.
[32] Stanley PE. Anal Biochem. 1971; 39: 441–453.
[33] Brolin SE, Borglund E, Tegnér L, Wettermark G. Anal Biochem 1972; 42: 124–135.
[34] Blinks JR. Volume II. In: VanDyke K, ed. Bioluminescence and Chemiluminescence; Instruments and Applications, Florida: CRC Press; 1985: 185–226.
[35] Stanley PE, Kelley BC, Tuovinen OH, Nicholas DJD. Anal Biochem. 1975; 67: 540–551.
[36] Smith DF, Stults NL, Rivera H, Gehle WD, Cummings RD, Cormier MJ. In: Stanley PE, Kricka LJ, eds. Bioluminescence & Chemiluminescence, Current Status. Chichester: John Wiley & Sons; 1991: 529–532.
[37] Meighen EA. In: Stanley PE, Kricka LJ, eds. Bioluminescence & Chemiluminescence, Current Status. Chichester: John Wiley & Sons; 1991: 3–10.

2 Principles of Analytical Applications of Luminescence

2.1 Compounds Which Can Be Analyzed

Applications of bioluminescence for analytical purposes usually involve a substance which plays an essential role for in vivo luminescence. We use the term "light inducer" to designate this compound. In the firefly, for example, ATP serves as the light inducer. In luminescent bacteria, it is $FMNH_2$ or NADH and NADPH, because of their ability to reduce FMN rapidly.

In the subsequent text we are using the symbol B to denote the bioluminescence inducer. The most straightforward analyses are measurements of B using an extract of luminescent organisms together with various additives. The extract consists primarily of a key enzyme, a luciferase L, and reactants R. Compounds which can be brought to participate stoichiometrically in a preceding B-producing reaction may be determined likewise. It is then a matter of analyzing substances via compound B. Obviously, it is also possible to use the scheme for the determination of the enzyme E, which is required by the preceding reaction. B is frequently an energy-rich compound, and the preceding reaction may be thought of as taking place between an energy donor D and an energy carrier C. This means that compounds of both types C and D can be determined (Table 2-1). The preceding B-producing reaction may be carried out as a prior, separate step, or allowed to proceed simultaneously with the bioluminescence reaction in a

Table 2-1. Principal schemes in bioluminescence analysis. – Determinable substances are: B Inducer of Bioluminescence, C Energy Carrier, D Energy Donor, E Coupling Enzyme, L Luciferase.

Reaction type	Substance analyzed (example)		
	B		L
(1) $(R +) B \xrightarrow{L}$ light	ATP		Firefly luciferase
	$FMNH_2$, NADH		Bacterial luciferase
Analysis of light inducer B (R symbolizes necessary reactants)			
	C	D	E
(2) $C + D \xrightarrow{E} B$	ADP	PEP	Pyruvate kinase
	$NADP^+$	G-6-P	G-6-P dehydrogenase
Analysis of nucleotides, substrates, and enzymes employing a reaction producing B (and analysis of B)			

common solution. The reaction scheme for the analysis might now be:

$$C + D \xrightarrow{E} B \qquad \text{(R 2-1)}$$

$$(R +)\ B \xrightarrow{L} \text{light} \qquad \text{(R 2-2)}$$

An example is the reaction between NAD^+ and sorbitol, which is catalyzed by sorbitol dehydrogenase. The substances to be determined are C NAD^+, D sorbitol and E sorbitol dehydrogenase.

2.2 Kinetic Considerations

There is an important difference between the application of luminescence and other photometric methods to follow the course of a chemical reaction. In luminescence, the light intensity is a direct measure of the reaction rate, whereas, for instance in fluorescence or absorption photometry, the intensity of the outcoming beam measures the concentration of a reactant or a product.

Consider the analytical situation in fluorescence analysis as compared with luminescence analysis of a light inducer B. In both cases an outgoing light signal is recorded as a result of an excitation. In the first case, the excitation is physical and in the second case, it is chemical (Figure 2-1).

To create physical excitation, incoming light with an intensity I_0 has to act upon B; and as a first approximation, the intensity of the emerging light I is directly proportional to I_0. A steady light signal is obtained. To induce chemical excitation, reactants R have to interact with B. In this case, however, B is consumed, and the light signal I varies with the time-course of the reaction, (cf. Figure 2-1). Increased luciferase activity affords higher initial emission and more rapid fading, but the total integral remains basically the same. Due to the more rapid consumption of B, emission curves of analyses with a highly active enzyme cross the curves from analyses with a less active enzyme. This does not happen, if the luciferase activity remains unchanged, and intensity differences are caused by high and low concentrations of B.

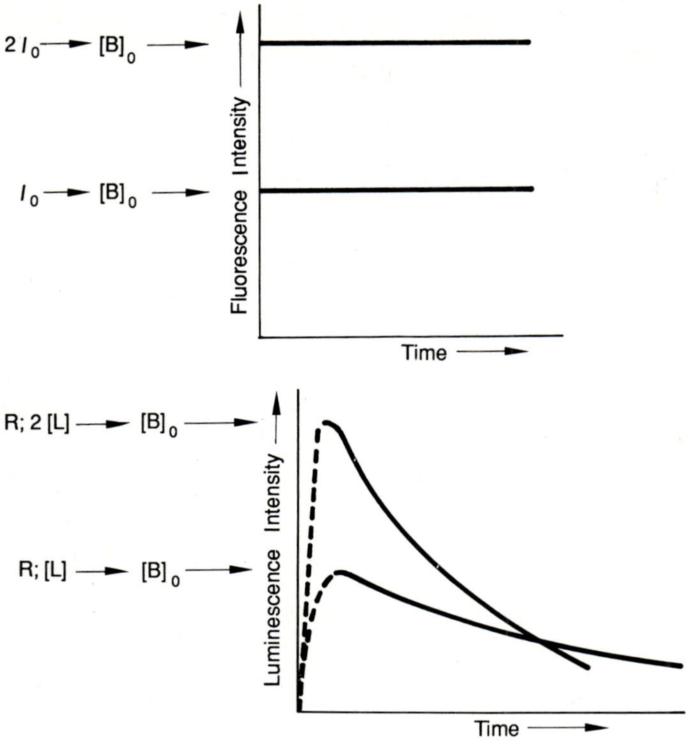

Figure 2-1. Schematic presentation of fluorescence analysis as compared with luminescence analysis of a compound B. In fluorescence analysis, a doubled intensity of the exciting light doubles the intensity of the ensuing fluorescence, which remains steady with a constant exciting beam. In fluorometry, the intensity is a measure of the *concentration*.
In luminescence analysis, B is consumed, and it is shown that doubling the luciferase concentration results in a twofold increase in initial maximal output and also in an increased fading rate, implying a crossing of the curves. There is usually a time lag before the maximal intensity is reached, which is indicated by the dotted initial parts of the curves. The intensity is a measure of the *rate* of the luminescence reaction, cf. equation (2-2).

2.3 Characteristics of the Light Emission

We shall now proceed to discuss some basic aspects of the kinetics of the analysis of a light-inducing compound B in a separate reaction. The analytical reaction is thus simply:

$$(R +)\ B \xrightarrow{L} \text{light.} \qquad (R\ 2\text{-}2)$$

Bioluminescence analysis is generally arranged with an excess of added reactants, R, allowing the concentration of luciferase, L, to determine the rate constant of the reaction. A surplus of oxygen is generally required and dissolved in the luciferase solution, from which it must not be removed. The light inducer B is kept at a very low concentration and therefore, as a first approximation, the reaction may be considered first-order with respect to B, i.e.

$$-d[B]/dt = k_2 [B] \tag{2-1}$$

k_2 being the rate constant of the overall luminescence reaction (R 2-2).

The intensity I of the emitted light becomes a measure of the reaction rate in being proportional to the decay rate of compound B

$$I = -b\ d[B]/dt, \tag{2-2}$$

where b is a constant. Consequently, the emission intensity I is proportional to the concentration of the light inducer B

$$I = b\ k_2 [B]. \tag{2-3}$$

Integration of the rate equation (2-1) yields the concentration of B:

$$[B] = [B]_0 \exp\{-k_2 t\}. \tag{2-4}$$

Accordingly, the emission intensity is described by

$$I = b\ k_2 [B]_0 \exp\{-k_2 t\}. \tag{2-5}$$

The emission intensity at a particular time is thus proportional to the initial concentration of B, i.e. for given values of t and k_2:

$$I_t = constant \cdot [B]_0. \tag{2-6}$$

This may be a better analytical parameter than estimates of initial rates, (cf. Figure 2-2.). Also, determining the amount of light (the light dose) A_t, i.e., integrating the light intensity over a fixed time leads to values that are linearly related to $[B]_0$, i.e., the initial concentration of B as

$$A_t = \int_0^t I\ dt = b\ [B]_0 (1 - \exp\{-k_2 t\}). \tag{2-7}$$

In other words, for fixed values of t and k_2, the amount of emitted light is

$$A_t = constant \cdot [B]_0. \tag{2-8}$$

Thus ensuring that each measurement proceeds with the same rate constant the integral over a fixed time may also be a good analytical parameter. The amount of light may either be recorded electronically; for instance through collecting the total number of photons emitted or, it may be obtained graphically by the integration of intensity versus time curves (cf. Figure 2-2).
The total integral of the light emission becomes

$$A_\infty = \int_0^\infty I\ dt = b\ [B]_0 \tag{2-9}$$

This value is not dependent on the rate of the reaction and should be a useful analytical parameter, but it is in many cases difficult to determine it correctly. The independence of the total integral of the rate of the bioluminescence reaction means that the amounts of light produced by a fast or a slow system ought to be essentially the

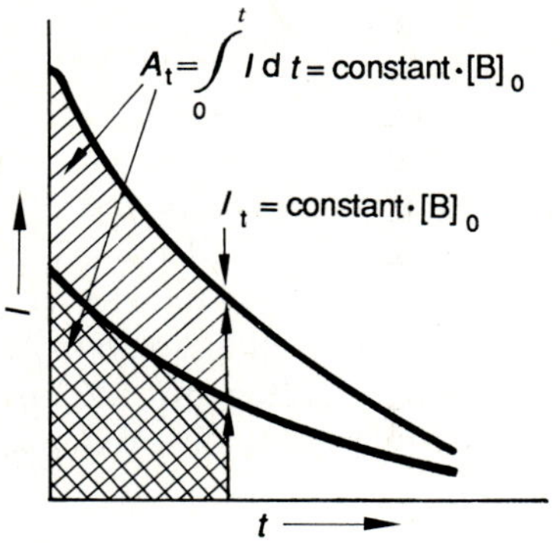

Figure 2-2. Fading curves, i.e., luminescence intensity versus time for an initial concentration $[B]_0$ and $2[B]_0$ of the light inducer B. The luminescence reaction is arranged as pseudo first-order, and it is assumed that the rate constant stays unaltered throughout the recordings. At any given time t, the intensity I_t is proportional to the initial concentration of B. This is also valid for the amount of emitted light A_t, represented by the dashed areas. The figure shows how doubling the initial concentration results in a twofold increase of emission intensity and light dose (A_t).

same. In particular this results in considerable differences in the light intensity level, depending on how the analytical system is set up. This is illustrated in Figure 2-3, which schematically shows differences in time profiles (light intensity versus time).

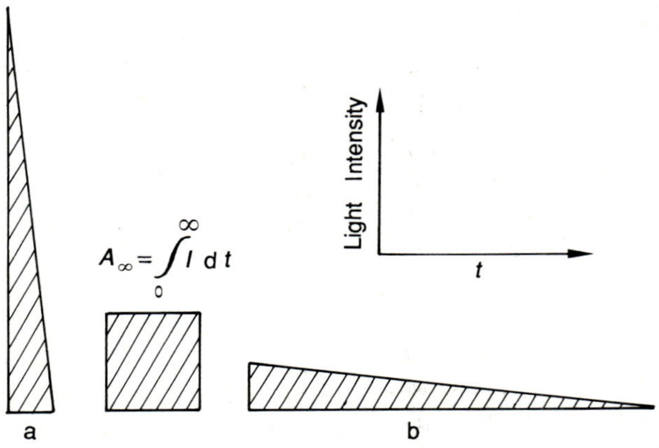

Figure 2-3. Principal variations of the time course of the emission in bioluminescence analysis; the ordinate denotes light intensity and the abscissa indicates time. For a given initial amount of a light inducer B, the available output (the total amount of emitted light, A_∞) stays the same. The amounts, as integrals of intensity versus time, are given by the areas of the respective figures. With an available emission of the size represented by the square, a fast reacting analytical system yields emissions appearing as a pulse of light (a), and a slowly reacting system affords a nearly steady light intensity for an extended time (b).

2.4 Assay Based on Products of a Preceding Reaction

We can now look at analyses based on the production of a light inducing compound by a separate preceding reaction, as exemplified in Table 2-1. Donors (D) as well as carriers (C) can be analyzed. If D is to be determined, an excess amount of C is maintained; and vice versa. The equilibrium position of the preceding enzyme reaction sometimes permits almost complete conversion of the target compound. This is often difficult, but it is in most cases unnecessary. Let us, for instance, assume that D is converted in a first-order reaction:

$$[D] = [D]_0 \exp\{-kt\}. \tag{2-10}$$

Since the formation of B is proportional to the consumption of D, we may write:

$$[B] = \mathit{constant} \cdot [D]_0 \tag{2-11}$$

for a given value of t.
Hence, it is sufficient to ensure that the rate constant and reaction time remain unchanged. For the analysis of nucleotides serving as carriers, $[C]_0$ replaces $[D]_0$ in equations (2-10) and (2-11). The scheme is likewise applicable to analyses of enzymes catalyzing the production of B. Here, an excess of C and D is maintained and the reaction is of zero-order, which means that the enzyme activity is proportional to the amount of B produced during a fixed reaction time.

The high sensitivity of bioluminescence assay, in fact, permits the use of a very limited extent of conversion in the preceding reaction, as well as in the subsequent luminescence reaction. The scheme is widely applicable, since many bioluminescence analyses can be carried out in two steps, as demonstrated for the coupling to bacterial luciferase [1]. The two step procedure is useful for the design of flow systems operating with immobilized enzymes in two compartments. In the first one, the light-inducing compound is formed and then carried to the second compartment, which contains luciferase and is located in front of a phototube [2].

References

[1] Brolin SE, Borglund E, Tegnér L, Wettermark G. Anal Biochem 1971; 42: 124–135.
[2] Roda A, Girotti S, Ghini S, Carrea G. J Biolum Chemilum. 1989; 4: 423–435.

3 Enzyme Kinetics

3.1 Particular Considerations of Enzyme Kinetics

3.1.1 Emission and Reaction Rate

In the preceding chapter, emphasis is placed on the fact that luminescence intensity is a direct indication of reaction rate. A simplified theoretical scheme is presented for designing analyses of bioluminescence-inducing compounds that are either present in a sample or formed in a separate preceding reaction. It is pointed out that none of the reactions must be carried to completion. Before proceeding with the design of bioluminescence analysis, we need to recall some basic enzyme kinetics of importance in this connection.

3.1.2 The Michaelis-Menten Relation

As a rule, enzymes employed in luminescence analysis can be assumed to follow Michaelis-Menten kinetics, according to the equation

$$-\mathrm{d}[\mathrm{S}]/\mathrm{d}t = k_{\mathrm{ES}}\,[\mathrm{E}]_0[\mathrm{S}]\,/\,(K_\mathrm{m} + [\mathrm{S}]) \qquad (3\text{-}1)$$

where [S] is the substrate concentration, k_{ES} the rate constant for the decomposition of the enzyme-substrate complex, $[\mathrm{E}]_0$ the enzyme concentration, and K_m the Michaelis constant. K_m has the unit of concentration, and if $K_\mathrm{m} = [\mathrm{S}]$, the reaction proceeds at half its maximal velocity.

The analytical setup is usually arranged such that the enzymes are practically saturated with all reagents except the substrate to be measured. It is customary to aim at concentrations which are 10 to 20 times higher than the K_m value; this corresponds to 91 and 95 % of the maximal rate.

In an analytical application, *when a substrate is to be measured,* it is useful to define the conditions such that linearity between reaction velocity and substrate concentration is obtained. As a general rule, it is sufficient to work at substrate concentrations well below the K_m value, $[\mathrm{S}] \ll K_\mathrm{m}$. Equation (3-1) is then reduced to a first-order equation:

$$-\mathrm{d}[\mathrm{S}]/\mathrm{d}t = [\mathrm{S}]\,k_{\mathrm{ES}}\,[\mathrm{E}]_0\,K_\mathrm{m}^{-1}. \qquad (3\text{-}2)$$

The first-order rate constant k_a thus includes enzyme activity and the Michaelis constant:

$$k_\mathrm{a} = k_{\mathrm{ES}}[\mathrm{E}]_0\,K_\mathrm{m}^{-1} \qquad (3\text{-}3)$$

In bioluminescence analysis, conditions are defined such that equation (3-2) holds true for luciferase, as well as for preceding enzyme reactions. If, for instance, a substrate D is analyzed according to the reactions in Table 2-1 through the formation of a light-inducing compound B, which subsequently serves as substrate in the luciferase reaction, linearity of the analytical response is obtained by using concentrations of D and B well below their respective K_m value. This approach is generally possible in luminescence analysis, as the high sensitivity permits working at concentrations well below K_m. An alternative analytical approach may be thought of, which allows reactions to proceed to completion. In this case, the substrate concentrations in relation to K_m are irrelevant and may be close to or exceed K_m. However, in practice it is difficult to develop useful analytical procedures based on complete conversions.

Likewise, *when analyzing enzyme activity,* it is common to saturate the system with the required substrates and cofactors, and equation (3-1) becomes

$$-d[S]/dt = k_{ES}\,[E]_0 \qquad (3\text{-}4)$$

A reaction of zero order is obtained, with a rate constant $k_{ES}[E]_0$

3.1.3 Inhibition of the Enzyme Reaction

For the measurement of a substrate S, when a competitive inhibitor U is present at a concentration [U], one obtains expression (3-5) for the first-order rate constant instead of (3-3), i.e.,

$$k_a = k_{ES}\,[E]_0\,K_m^{-1}\,(1 + [U]/K_u)^{-1} \qquad (3\text{-}5)$$

where K_u is the inhibitor constant. Inhibitors of the luminescence reaction may be produced in the reaction or be present originally. If inhibitors are produced in the reaction, this should perturb the exponential decay and *increase* the rate of fading during the early part of the emission. Energy is preserved, resulting in an elevated tailing-out of the emission as compared with the uninhibited case, see Figure 3-1.

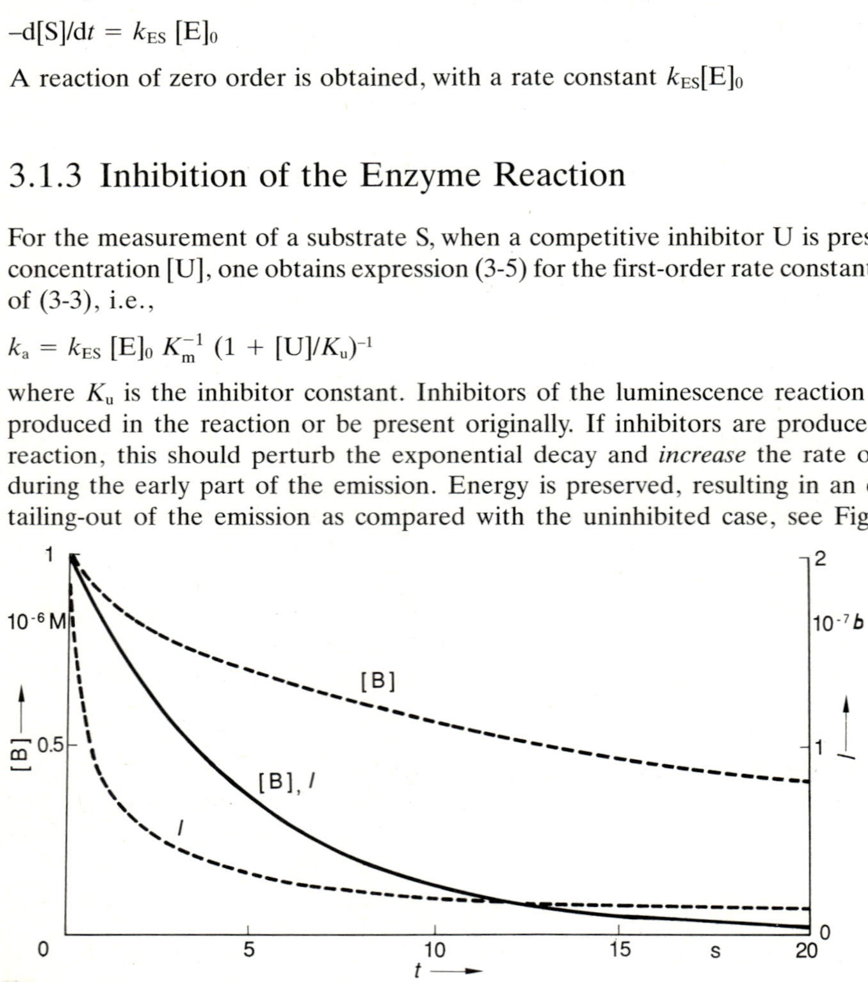

Figure 3-1. Effect of product inhibition during the analysis of the substrate B, which takes part in the luminescence reaction:

B → light + U (R 3-1)

It is, accordingly, assumed that U is formed as a product of the reaction and is not present originally. Luminescence curves are shown both for non-interference of U with the luminescence reaction and for its role as an inhibitor. Two ordinates are given: concentration of B and intensity, respectively.

Solid line: Product U causes no inhibition. The conditions are such that the first-order rate constant of equation (3-3) applies.
Broken line: U is a competitive inhibitor, and equation (3-5) applies.
The following assumptions are made: Initial concentration of B, $[B]_0 = 10^{-6}$ M; k_{ES} $[E]_0$ K_m^{-1} = 0.2 s^{-1}; $K_u = 10^{-7}$ M. The luminescence intensity is assumed to be proportional to the rate of the decay of B, with the proportionality constant b, i.e., $I = -b$ (d[B]/dt).

Addition of inhibitors may provide a means to reduce the reaction rate and is one of the principles underlying the nearly-constant emission of the slowly reacting firefly-luciferase system [1, 2].

3.1.4 Enzyme Activities and Rate Constants

Enzyme activities are usually expressed in terms of units U. One unit indicates the capability of converting 1 μmole of substrate per minute.

The specific activity is defined as the activity (expressed in units) per mg of enzyme. Thus,
1 U = 1 μmole substrate min^{-1} = 1.67×10^{-8} mole s^{-1}.
The definition refers to a situation in which the conversion is not limited by the substrate concentration and may thus be described by equation (3-4), and we obtain:

k_{ES} specific activity (U mg^{-1} enzyme)
$[E]_0$ concentration of enzyme (mg L^{-1})
k_{ES} $[E]_0$activity per unit volume (U L^{-1}; M s^{-1})
(1 U L^{-1} = $1.67 \cdot 10^{-8}$ M s^{-1})

The definitions given above may be valid over a limited range of enzyme concentrations. In particular, at high enzyme concentrations, the specific activity may decrease, as the approximations that the equations are based on cease to apply.

3.2 Selecting Analytical Parameter when Determining a Light-Inducing Compound

3.2.1 Measuring the Intensity Level or Integrating the Curve of the Emitted Light

When employing reaction (R 2-2):

(R +) B → light, (R 2-2)

for determining the compound B, the choice of analytical approach is normally between measuring the peak height and measuring the integral of the emitted light. Both methods have their advantages and disadvantages. In particular, there are some innate problems connected with the kinetics of the process, which will be discussed below. As indicated in Chapter 2, it is in principle possible to use the intensity at any given reaction time as a basis for drawing up calibration curves. There is a high probability that a linear relation exists between the intensity and the amount of substance to be analyzed. In practice, however, it may be difficult to reproduce mixing times etc. and to select an arbitrary time for recording the light signal. This is a source of errors. It presupposes, in fact, that the intensity does not vary very much during the recording.

Current analytical methods are usually based upon either fast or very slowly proceeding luciferase reactions. With fast reacting luciferase, the portion of the entire flash profile that is recorded may vary. The nature of the curve, with a steep rise and a rapid decay, is such that intensity recordings easily become ill-defined, unless they are carried out at maximal intensity. Selecting this maximum is therefore common under such circumstances. With a slowly reacting luciferase, observing a well-defined reaction time or recording the maximal light emission is far less critical. In fact, maximal emission may occur so early during the recording that it is not picked up by the measuring equipment, and yet the recording may be satisfactory.

Integrating the light signal may also be a suitable approach, resulting in a linear relation between the substance to be analyzed and the integral, cf. above. As with a fast reacting luciferase, essentially the entire flash profile is obtained. In this case, it may seem attractive to measure the total integral, i.e.

$$A_\infty = \int_0^\infty I \, dt \; . \tag{3-6}$$

However, as the decay is exponential, this requires long integration times and introduces uncertainties, due to background light and side reactions. It should be recalled that a major advantage of using fast reacting luciferase is the possibility to suppress influences from background light by keeping reaction times short and obtaining a high signal level. This implies also that the integration time should be kept short. Usually, integration is carried out over a fixed time after the start of the reaction or after the maximum, covering approximately 80–90 % of the total integral.

3.2.2 Rate Requirements for Adequate Sample Handling

The handling of the samples and the intensity determinations are carried out within a finite time. Of particular importance is the time required for starting the reaction by mixing samples and reagents. This presents difficulties in defining zero time. Another difficulty may arise from trying to measure a changing intensity at a well-defined time. It may therefore be of interest to obtain an estimate of possible errors in the recording of the intensity that are based on uncertainties in the time scale. The relative error in the determination of emission intensity is given by

$$\Delta I/I = k_2 \, \Delta t, \tag{3-7}$$

if equations (3-2) and (3-3) are valid for reaction (R 2-2), i.e., the luminescence reaction. The expression should thus be a first approximation of the error in the analysis of a light-producer B.

Manual handling of the sample may bring about an uncertainty of ± 5 s in the time scale. Accepting an error of 1 % in the intensity determination and using equation (3-7) one may arrive at the conclusion that reaction (R 2-2) may in this case be allowed to proceed with a rate constant of up to 10^{-3} s^{-1}. To be able to use a faster luminescence reaction and to obtain still higher light intensities, it is necessary to employ experimental procedures other than the manual mixing of samples. With stopped flow techniques, a Δt of 10 ms may be achieved fairly easily. Accepting this value and 1 % error in intensity, one obtains $k_2 = 1$ s^{-1}, permitting a very fast luminescence reaction in the analysis. Such a high value is, however, hardly realistic for analytical applications.

The great advantage of being able to employ a fast luminescence reaction should be appreciated. This is basically due to the accompanying increase in emission intensity, I, which is proportional to k_2 in equation (2-3). The higher intensity often facilitates reaching higher sensitivity levels. For the cases presented in Table 3-1, there is thus also a

Table 3-1. Permitted rates of the luminescence reaction, according to various assumptions with respect to uncertainties in the time measurements and systematic errors in the determination of emission intensity.

Analytical situation	Accepted error in intensity determination	Uncertainty in time measurement	Maximal value of rate constant of luminescence, k_2
Manual mixing	1 %	10 s	10^{-3} s^{-1}
Stopped flow	1 %	10 ms	1 s^{-1}

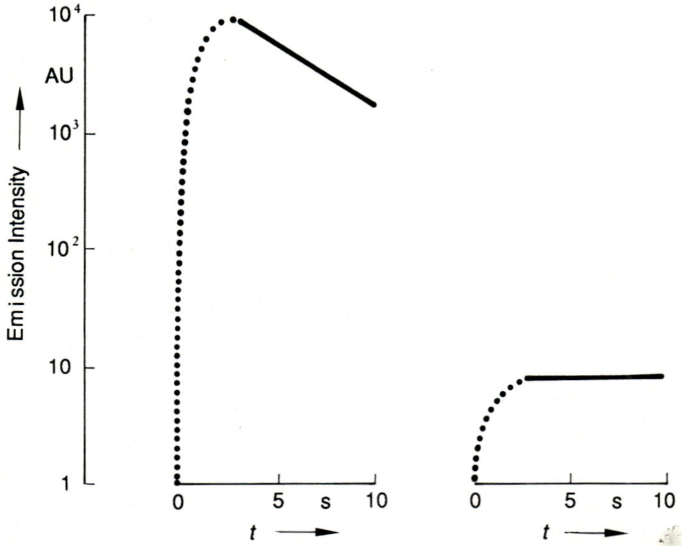

Figure 3-2. Comparison of emission intensities (given in arbitrary units) with fast and slowly reacting luciferase. For the fast system, a rate constant of 0.27 s^{-1} is assumed, corresponding to about 25 % degradation per second. For the slow system, the assumed rate constant is 1.7×10^{-4} s^{-1}, with 1 % degradation per minute.

thousandfold difference in initial light intensity. Slowly reacting analytical systems are often arranged such that no more than 1 % of the substrate is converted per minute. This corresponds to a rate constant of 1.7×10^{-4} s^{-1}, which is much lower than the permitted value for manual mixing in Table 3-1.

Typical analytical approaches using fast or slowly reacting luciferase are compared in Figure 3-2. The light flux is calculated, using conversion rates which are valid in the analysis of NADH with the bacterial luciferase system. The first case is relevant for a system designed for injection or stopped-flow techniques [3, 4]. This allows reaching a very high sensitivity. The second case refers to the choice made by commercial kit producers and permits relatively slow mixing of reactants and the substance to be measured, which facilitates the analytical procedure [1, 2, 5]. Further additions of internal standards and reactants during a measurement are also possible with this technique, due to the slow fading. Analytical situations may arise which make it useful to work in an intermediate rate region. This matter is discussed in the following chapters.

3.2.3 Representation of the Start-Up Phase

The simplified theory of luminescence reactions in analysis, as outlined in Chapter 2, does ot account for delays in the start-up phase of the reaction. After the reagents that induce the emission of light have been mixed, one must expect a certain delay before the intensity begins to increase. In practice, a fast reacting luminescence reagent system yields emission curves with a clear maximum, followed by a rapid fading; whereas with a slowly reacting system, the maximum is less well defined, but followed by a long-lasting intensity decline.

In particular, reaction (R 2-2) is a gross simplification and involves a considerable number of reaction steps. This is evident when we compare the simplified presentation of the firefly reaction with the complicated chain it represents, as described in Chapter 5. There are also delays in the start-up phase of the reaction, due to the mixing process itself and to the binding between substrates and enzymes, etc. The result is a non-instantaneous build-up of emission intensity, even in cases where the simple theory predicts that the emission should reach maximum intensity in the beginning.

This start-up phase, though in most cases very complicated, may be represented as an activation process operating in the reaction mixture prior to the emission. At a first approximation, this start of the process can be considered a first-order reaction with a rate constant k_{st}. The luminescence reaction is regarded as a consecutive process. For the case of the luciferase reaction, one obtains

B → B′ first-order rate constant: k_{st}
(R +) B′ → light first-order rate constant: k_2.
It follows that:

$$I = [B]_0\, b\, k_2\, k_{st}(k_2 - k_{st})^{-1}\, (\exp\{-k_{st}t\} - \exp\{-k_2t\}) \quad (3\text{-}8)$$

where b is the proportionality factor between reaction rate and emission intensity, cf. equation (2-2).

Using expression (3-8) to simulate emission curves of the light-producing substance B may help to explain some important properties of the intensity and the integral of the intensity curve as analytical parameters. This is illustrated in Figure 3-3, which lists values for the emission that are representative of fast reacting luciferase. The ordinate indicates the emission intensity I in units of $b\,[B]_0$ (b times initial concentration of B). It is assumed that the build-up proceeds with a rate constant which is 10 times higher than that of the decay. For comparison, a dotted curve indicates instantaneous build-up (k_{st} → infinity), which results in the curves predicted by equation (2-5). By introducing an expression for the build-up phase, the maximum is shifted to a few seconds after the start of the process and is accompanied by a somewhat lower maximum than in cases in which no build-up time is required (77 %). From then on, the time course of the emission is affected very little by variations in the assumed build-up. For some reaction times, the value of the integral is indicated on the graph for some reaction times relative to the total integral with the value $b\,[B]_0$ It is shown that in following the reaction for 30 s, which seemingly includes the essence of the flash profile, one still neglects 5 % of the total integral.

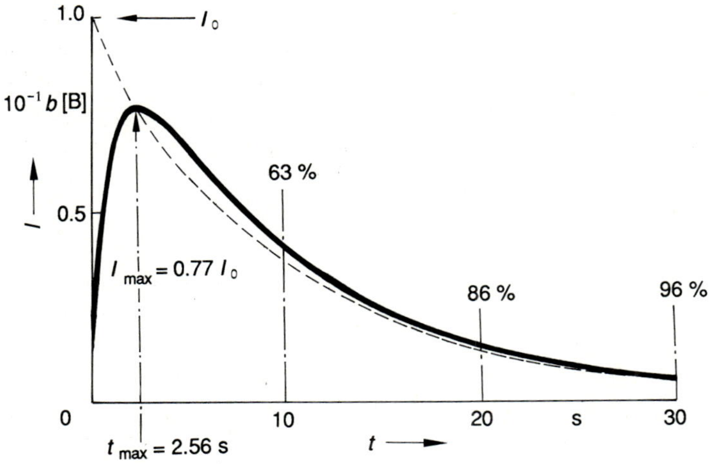

Figure 3-3. The influence of a lag time, illustrated by simulated emission curves according to equation (3-8). It is assumed that the luminescence reaction proceeds with a rate constant k_2 = 0.1 s^{-1}. The solid curve represents the emission when the build-up proceeds with k_{st} = 1 s^{-1}. The dashed curve illustrates the emission when the build-up is infinitely fast, i.e., $k_{st} \rightarrow \infty$; Hence, equation (3-8) is reduced to that of a normal exponential decay, equation (2-5).
It is shown that the assumed delay in the build-up causes the maximum emission to be only 77 % of its theoretical value following an infinitely fast formation process. The percentages show the extent to which the total emission is reached for the reaction times indicated, i.e., A_t/A_∞, where

$$A_t = \int_0^t I\,dt \qquad (3\text{-}9)$$

In a given start-up process, the intensity maximum is proportional to the initial concentration of B. In fact, this holds true for intensity readings at any given time.

The occurrence of a build-up phase with the kinetics described above will thus be compatible with selecting the intensity value or the integral at any reaction time as a quantity that is linearly related to the concentration of the substance to be analyzed. Neglecting the build-up affords the same proportionality. Hence, we regard the simplified scheme presented in Section 2.3 as suitable for designing bioluminescence analyses.

In the next chapter we will enter into schemes with simultaneously proceeding coupled reactions. Lag times, a common feature of enzyme reactions, have to be expected. However, it is believed that the consideration of a start-up process would add little to the general conclusions obtained from equations where influence from the start-up process is neglected. Calibrations with known amounts of substrate or of the target enzyme are advisable and should reveal any significant deviations from the predicted behavior.

Nevertheless, certain possible divergences may deserve some comment with regard to the two analytical alternatives of initial rate measurements or steady-state recordings. Lag times will naturally entail the recording of a somewhat slower initial emission rate than the calculated one, but not one which may give rise to deviations from the proportionality between rate and analyte concentration (or in enzyme assay, between rate and enzyme activity). A slightly retarded onset means that the time for approaching the calculated maximum velocity will be prolonged. This is easily compensated for by allowing a longer recording time to reach the steady-state. The consumption of energy during a prolonged initial emission lowers the subsequent supply and may result in a reduced steady-state level; but proportionality may still be obtained.

References

[1] Lundin A. In: Serio M, Pazzagli M, eds. Luminescent Assays Perspectives in Endocrinology and Clinical Chemistry. New York: Raven Press; 1982: 29–45.
[2] Wulff K, Haar H-P, Michael G. In: Serio M, Pazzagli M, eds. Luminescent Assays Perspectives in Endocrinology and Clinical Chemistry. New York: Raven Press; 1982: 47–52.
[3] Wettermark G, Brolin SE, Hjertén S. Cell Mol Biol 1977; 22: 329–339.
[4] Brolin SE, Hjertén S. Mol Cell Biochem. 1977; 17: 61–73.
[5] McElroy WD, DeLuca M. In: DeLuca MA, McElroy WD, eds. Bioluminescence and Chemiluminescence Basic Chemistry and Analytical Applications. New York: Academic Press; 1981: 179–186.

4 Coupling of Reactions

4.1 Principal Approach

Many compounds can be determined by two consecutive reactions. The analytical situation discussed in Section 2.4 refers to the case when these reactions are carried out separately, i.e., when the light inducing compound B is formed in the first reaction, which is stopped before the luminescence reaction is started. In other situations, which will be discussed in this chapter, it may be convenient to couple reactions in such a way that B is formed as it is consumed in the luminescence reaction [1–4]. In this case, the reactions proceed together in the same solution.

Using the symbols defined in Table 2-1, we arrive at the following scheme.

Reaction producing compound B:

$$C + D \xrightarrow{E} B \qquad \text{(R 4-1)}$$

Luminescence reaction consuming light-inducing compound B:

$$(R +)\ B \xrightarrow{L} \text{light} \qquad \text{(R 4-2)}$$

The light-inducing compound may, for instance, be ATP, NAD(P)H, or $FMNH_2$. Since the reactions take place simultaneously in the same solution, the end products of the luminescence reaction may enter into the equation for reaction (R 4-1). This may give rise to an enzymatic cycle, in the case when compound C is a product of the luminescence reaction:

$$(R +)\ B \xrightarrow{L} \text{light} + C \qquad \text{(R 4-2')}$$

We shall first treat the non-cycling case and later discuss situations in which cycling is important. Measurable species are the enzyme E, the energy carrier C, and the energy donor D. Examples of analytical applications are given in Table 4-1.

Table 4-1. Assays using simultaneous coupled reactions

Type		Example
Without cycling:		
C + D → B	(R 4-1)	ADP + PEP → ATP
(R +) B → light	(R 4-2)	(R +) ATP → light (no ADP recovery)
With cycling:		
C + D → B	(R 4-1)	$NADP^+$ + (G-6-P) → NADPH
(R +) B → light + C	(R 4-2′)	(R +) NADPH → light + $NADP^+$ (recycled)

4.2 Non-Cycling Coupled Reactions in Substrate Analysis

4.2.1 General Behavior

The analysis of a compound C [5, 6] or D [7-9] through reaction (R 4-1), coupled to the conversion of the light inducing product B in the luminescence reaction (R 4-2), is usually arranged such that it may be treated as consisting of two simultaneous first-order reactions. Consequently, during the analysis of D, an excess of C is maintained, and vice versa. The pseudo first-order rate constants for the two steps are k_1 and k_2, respectively. Based on the premises mentioned above, we shall assume that the substrate to be analyzed is D, with initial concentration $[D]_o$. If, instead, C is analyzed, the concentration of C replaces that of D in the following equations. Taking into account that the formation of B runs parallel with its consumption, its concentration at time t follows the equation:

$$[B] = [D]_0\, k_1(k_2-k_1)^{-1}\, (\exp\{-k_1t\} - \exp\{-k_2t\}). \qquad (4\text{-}1)$$

As discussed in Chapter 2, the luminescence intensity I can be considered to be proportional to the rate of reaction (R 4-2) and is thus proportional to the concentration of B:

$$I = b\, k_2\, [B]. \qquad (4\text{-}2)$$

b is the same proportionality constant as the one defined in Chapter 2, in connection with the determination of a light inducer B. It describes the proportionality between the rate of reaction (R 4-2) and emission intensity. Combining equation (4-1) with equation (4-2) results in the equation:

$$I = b\, k_2\, [D]_0\, k_1(k_2-k_1)^{-1}\, (\exp\{-k_1t\} - \exp\{-k_2t\}). \qquad (4\text{-}3)$$

One obtains curves with the emission intensity rising to a maximum whose height depends on the initial concentration of the compound to be measured, $[D]_0$, as well as on the value of the rate constants k_1 and k_2.

The reaction time that is required to reach the maximum is independent of the initial concentration of D and is given by

$$t_{max} = (k_2-k_1)^{-1}\ln(k_2/k_1). \qquad (4\text{-}4)$$

The height of the maximum is then:

$$I_{max} = b\, k_2\, [D]_0 \cdot \left(\frac{k_2}{k_1}\right)^{\frac{k_2}{k_1-k_2}}\ . \qquad (4\text{-}5)$$

Figure 4-1 illustrates the way the intensity I varies during the course of the reaction, if k_1 and k_2 are of the same order of magnitude.

From equation (4-2) it is evident that if the conditions of the analysis can be kept sufficiently reproducible as far as the constants b, k_1 and k_2 are concerned, it is possible to select a time and to use either intensity measurements or the integral as a parameter which is proportional to the concentration of the compound to be measured, i.e.

$I_t(\text{fixed } k_1, k_2) = constant_1 \cdot [D]_0.$ (4-6)

and

$$\int_0^t I(\text{fixed } k_1, k_2)\, dt = constant_2 \cdot [D]_0. \quad (4\text{-}7)$$

In particular, the height of the maximum is proportional to $[D]_0$. Recordings of the peak height may thus serve as a suitable measure of the initial concentration of D.

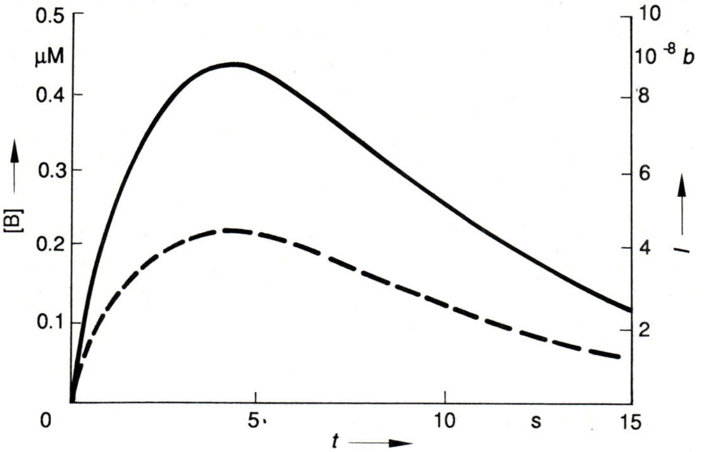

Figure 4-1. The time course of the emission intensity when using two simultaneous coupled reactions for the assay of a compound D. The two reactions are arranged to become first-order with respect to B and D. A case is illustrated in which the two rate constants are not identical, but of the same order of magnitude. The reaction diagram is:

$C + D \xrightarrow{E} B$ rate constant $k_1 = 0.3\ s^{-1}$

$R + B \xrightarrow{L} \text{light}$ rate constant $k_2 = 0.2\ s^{-1}$

Curves for two different initial concentrations of D are shown:

—— $[D]_0 = 1 \times 10^{-6}$ M; - - - - $[D]_0 = 0.5 \times 10^{-6}$ M.

The left ordinate scale gives the concentration of B and the right scale the intensity readings. The intensity is directly proportional to the rate at which B is consumed through luminescence, and since this reaction is arranged to become first-order in B, the proportionality also holds for the concentration of B, cf. equation (4-2).
In assays, the emission is usually recorded at a selected time t, for instance at a maximum, and it is proportional to the initial concentration of D, cf. equation (4-6). It is also possible to use integrations and to measure the emitted light dose, equation (4-7).

In principle, it is possible to arrange the reactions in three different ways (Table 4-2). In general, higher values of k_1 and k_2 imply higher emission intensities and higher sensitivities, but obviously there is a limit to the possibilities of increasing the rate constants. Selecting about the same value for k_1 and k_2 may be advantageous for high

Table 4-2. Choice of rate constants for analytical applications.

	Advantages	Disadvantages
$k_1 \simeq k_2$	High sensitivity	Stability of all enzymes is crucial
$k_1 << k_2$	Early maximum and nearly-constant light emission; low demands on luciferase stability	Stability of enzymes in reaction (R 4-1) is crucial
$k_1 >> k_2$	Early maximum and nearly-constant light emission; low demands on stability of reaction (R 4-1) enzymes	Stability of luciferase is crucial

degrees of sensitivity, as the entire emission comes out as a pulse, which can be integrated or whose maximum may be determined. The symmetry of the rate constants k_1 and k_2 in the equations (4-4) and (4-5) suggests that, if either reaction (R 4-1) or reaction (R 4-2) is rate-determining, the same type of emission curve should be obtained for both cases. In analytical practice, it is common to let k_1 and k_2 be of different orders of magnitude, which simplifies the expressions for [B] and I, which are discussed below.

4.2.2 Defining a Rate-Determining Reaction

If k_1 and k_2 are very different, the maximum of the intensity curve shifts towards the beginning of the emission, and the bulk part of the emission is determined by the slower reaction. The intensity and the time course of the fading are therefore determined by the rate of the preceding enzyme reaction, if the luminescence reaction is faster; and by the luminescence reaction, if it is slower. If the luminescence reaction is faster, equation (4-3) may be written:

$$I \simeq [\mathrm{D}]_0\, b\, k_1 \exp\{-k_1 t\} \text{ for } k_1 << k_2 \tag{4-8a}$$

and if the luminescence reaction is slower, it follows that:

$$I \simeq [\mathrm{D}]_0\, b\, k_2 \exp\{-k_2 t\} \text{ for } k_1 >> k_2. \tag{4-8b}$$

These equations are analogous to that for the analysis of a light-inducing compound B, equation (2-5), but in this case the light intensity is determined by the concentration $[\mathrm{D}]_0$ and by the rate constant of the slower of the two consecutive steps. The luminescence intensity approaches a maximum value:

$$I \simeq [\mathrm{D}]_0\, b\, k_1 \tag{4-9a}$$

and

$$I \simeq [\mathrm{D}]_0\, b\, k_2, \tag{4-9b}$$

respectively. Practically, this implies that with faster luminescence reaction, the intensity remains essentially independent of changes in luciferase activity (cf. Figure 4-2). Correspondingly, changes in the activity of the coupling enzyme may be

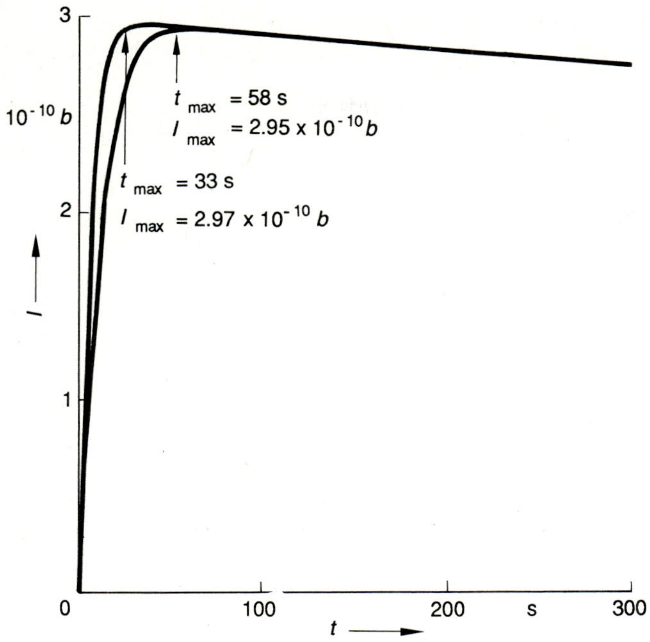

Figure 4-2. The initial light emission from an analytical scheme based on two consecutive first-order reactions D → B → light, with a fast luminescence reaction preceded by a very much slower enzyme reaction. The effect of variation in luciferase activity is illustrated.
It is shown that the decay curve stays unaltered, even with a drastic change in luciferase activity; and that virtually the same luminescence level and maximum intensity are reached. Calculations from equations (4-3) to (4-5).
The scheme illustrated is:

$C + D \xrightarrow{E} B$ rate constant:

$k_1 = 0.0003\ s^{-1}$

$B + R \xrightarrow{L} light$ rate constant:

$k_2 = 0.2\ s^{-1}$ (left curve)
$k_2 = 0.1\ s^{-1}$ (right curve)

D is analyzed and its initial concentration is in this case $[D]_0 = 10^{-6}$ M.

tolerated if the luminescence reaction is slower and if it proceeds at an unimpaired rate. This may be the case with immobilized luciferase.

The initial build-up of intensity is essentially the faster reaction. Thus, if the luminescence reaction is faster, an exponential intensity increase is determined by the rate constant k_2 of the luminescence reaction:

$$I \simeq [D]_0\, b\, k_1\, [1 - \exp\{-k_2 t\}]. \qquad (4\text{-}10)$$

In analytical practice, it is often a matter of compromise between sensitivity, in terms of satisfactory intensity, and reproducibility, mainly in terms of avoiding disturbances in enzyme activity. Taking into consideration the required mixing times is also of importance. In the example chosen in Figure 4-2, the luciferase activity has changed considerably, since it is reduced by half; but the intensity maximum is only reduced by less than 1%. Increasing the rate of the coupling reaction, by enlarging k_1 raises the intensity and results in a higher sensitivity. The setup will, however, be more sensitive to changes in luciferase activity, if k_1 approaches k_2.

4.2.3 Analysis Using Initial Rates

The initial phase of the luminescence is dependent on the rate of both reactions, described by

$$(\mathrm{d}I/\mathrm{d}t)_0 = [\mathrm{D}]_0\, b\, k_1\, k_2. \tag{4-11}$$

Attempts have been made to determine $[\mathrm{D}]_0$ by initial rates; but it is difficult to arrive at suitable conditions for such analyses. Primarily, it is necessary that the two reactions (R 4-1) and (R 4-2) proceed unimpaired. Furthermore, any disturbances or deviations from the ideal in the start-up process can easily distort the results.

4.3 Cycling Reactions in Substrate Analysis

4.3.1 The Steady-State

In the previous sections, analytical reactions were described in which an energy donor D or an energy carrier C were determined by two simultaneous reactions (R 4-1) and (R 4-2), and it was presupposed that a product of reaction (R 4-2) was not a substrate of reaction (R 4-1). However, it is often possible to arrange an analytical situation such that compound C is regained in reaction (R 4-2), cf. Table 4-1 and Figure 4-3. This has

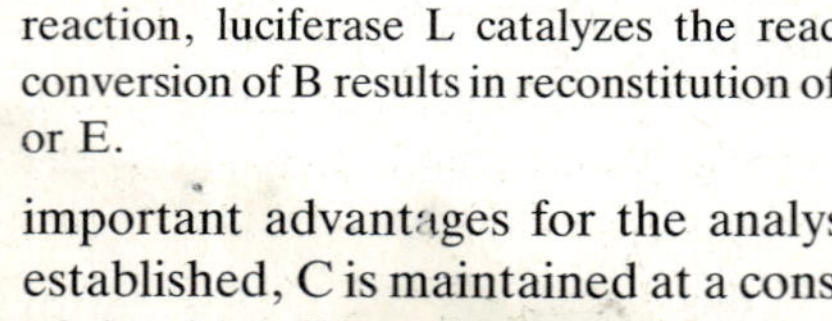

Figure 4-3. A useful enzymatic cycle in luminescence assay. In a reaction catalyzed by enzyme E, an energy donor D reacts with an energy carrier C to form the light producer B. In the luminescence reaction, luciferase L catalyzes the reaction of reactants R with B. A cycle is closed, if the conversion of B results in reconstitution of C. The scheme can be employed for the analysis of C, D, or E.

important advantages for the analysis of C, because once the equilibrium has been established, C is maintained at a constant level. Another advantage may be a reduction of the side effects due to unwanted reactions. While a steady-state is approached, any parallel reactions producing B will essentially self-terminate, due to the lack of substrate. Only the desired analytical reaction is specifically supplied with excess substrate. This may, for instance, be useful in the determination of either NAD(H) [5] or NADP(H) [6], if both are present in the same sample.

The reaction scheme for the cycle may now be written

$$C + D \xrightarrow{E} B \qquad \text{rate constant } k_1 \tag{R 4-1}$$

$$(R +)\ B \xrightarrow{L} \text{light} + C \quad \text{rate constant } k_2 \tag{R 4-2'}$$

Again, the analytical conditions are usually arranged such that the reactions become first-order with respect to C and B, respectively. A basic prerequisite is an excess of D and R. As before, the light intensity I is determined by the rate of reaction (R 4-2′), and it is proportional to the concentration of B, according to equation (4-2). The intensity approaches equilibrium, according to the equation:

$$I = [C]_0\, b\, k_1\, k_2(k_1 + k_2)^{-1}\, [1 - \exp\{-(k_1 + k_2)t\}]. \tag{4-12}$$

It should be noticed that at any time t, the intensity is proportional to the initial concentration of C. The intensity at a fixed time may thus be used as a measure of $[C]_0$, but it is analytically more attractive to wait for the constant light intensity produced by the cycle i.e., the equilibrium light intensity given by

$$I_\infty = [C]_0\, b\, k_1\, k_2/(k_1 + k_2). \tag{4-13}$$

The intensity can obviously be increased by selecting systems with high rate constants k_1 and k_2. If the values of the rate constants differ considerably, the height of the equilibrium intensity is determined by the rate constant of the slower step, whereas the approach to the equilibrium is determined by the rate constant of the faster step. Hence in case $k_2 >> k_1$,

$$I = [C]_0\, b\, k_1\, [1 - \exp\{-k_2 t\}] \tag{4-14}$$

and

$$I_\infty = [C]_0\, b\, k_1, \tag{4-15}$$

as exemplified in Figure 4-4.

The level of the equilibrium intensity is a convenient analytical parameter. Equation (4-12) makes it possible to estimate the time required to get sufficiently close to the asymptotically approached steady-state – by calculating the numerical value of the term $\exp\{-(k_1 + k_2)t\}$. Thus, if, for instance, the value of the term is 0.02, we have reached 98 % of the steady-state intensity. It is obvious that the analytical conditions have to be arranged such that at least one of the reactions (R 4-1) or (R 4-2′) proceeds fast, in order to reach a satisfactory intensity for measurement within a reasonable time.

Depending on the relative size of the rate constants: namely either $k_1 \simeq k_2$, $k_1 << k_2$, or $k_1 >> k_2$, one obtains the same properties in terms of dependence on enzyme and luciferase stability as indicated in Table 4-2. The advantage of a fast reacting luciferase system may be of particular importance, since partial inactivation of luciferase is tolerable, as exemplified in Figure 4-4. For measurements of small quantities of C, it

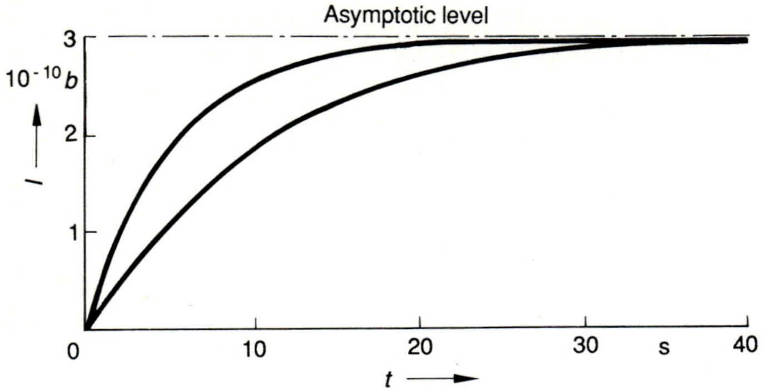

Figure 4-4. Analysis of C through enzymatic cycles, cf. Figure 4-3 and equation (4-12). Effect of variations in luciferase activity is shown for a system in which the luciferase reaction is the much faster step. The same steady-state light intensity is approached.
The two reactions of the scheme are arranged to become first-order with respect to B and to the measured compound C:

$$C + D \xrightarrow{E} B \qquad \text{rate constant:} \quad k_1 = 0.0003\ \text{s}^{-1}$$

$$R + B \xrightarrow{L} C + \text{light} \qquad \text{rate constant:}$$

$$k_2 = 0.2\ \text{s}^{-1}\ \text{(upper curve)}$$
$$k_2 = 0.1\ \text{s}^{-1}\ \text{(lower curve)}$$

The initial concentration of C is assumed to be $[C]_0 = 10^{-6}$ M.
In a setup for a series of assays, the intensity is recorded after reaching steady-state and is proportional to the initial concentration of C, cf. equation (4-13). Integration of the emission may lead to higher sensitivities.

may be desirable to raise the cycling rate and the equilibrium intensity level defined by the overall constant $k_1 k_2/(k_1 + k_2)$, cf. equation (4-13). If k_1 and k_2 assume very different values, the steady-state is insignificantly affected by large changes in the rate of the fast reaction. Figure 4-4 and Table 4-3 provide a basis for discussing the effects of variations in the analytical parameters, in situations when the luciferase reaction is considerably faster than the preceding reaction, $k_2 >> k_1$. Even drastic reductions of luciferase activity result in only small changes of the steady-state level but an impaired luciferase activity means a longer delay in approaching this level. However, as indicated in Table 4-3, the parameters can be chosen such that this time lag is short, in all cases less than a minute for reaching 99 % of steady-state intensity, and should therefore be unproblematic.

Attempting to increase sensitivity by letting the rate constant of the preceding enzyme reaction approach that of the luciferase reaction results in a greater dependency on the luciferase stability. When $k_1 \simeq k_2$, there is no compensation for decreased luciferase activity, and internal standards have to be used. In a specific analytical situation, it may be favorable to compromise between reduced but still adequate compensation for changes in luciferase activity and intensity loss.

Table 4-3. Effects of decrease in luciferase activity on the analytical results when using fast reacting luciferase ($k_2 = 0.2\ s^{-1}$) in coupled reaction cycles. The steady-state intensity I_∞ according to equation (4-15) is assumed to represent the sensitivity and may be expressed by the rate constant k_1.
Using equation (4-13) it is seen that I_∞ is only slightly affected by a decrease in luciferase activity, except at high sensitivities, when the enzyme of reaction (R 4-1) approaches the luciferase activity. In the last example, the overall rate is such that a cycle is completed within one minute. A less drastic decrease in luciferase activity than assumed in the table may allow satisfactory analytical accuracy. In all examples, the time needed to approach the steady-state is acceptable.

Sensitivity, k_1	Relative sensitivity	Decrease in I_∞ at halved luciferase activity ($k_2 = 0.2 \rightarrow 0.1\ s^{-1}$)	Reaction time t to reach 99 % of I_∞, assuming $k_2 = 0.2$	$k_2 = 0.1\ s^{-1}$
$3\times10^{-4}\ s^{-1}$	1	<0.2 %	23 s	46 s
$6\times10^{-4}\ s^{-1}$	2	<0.3 %	23 s	46 s
$1.5\times10^{-3}\ s^{-1}$	5	<0.7 %	23 s	45 s
$3\times10^{-3}\ s^{-1}$	10	<1.5 %	22 s	44 s
$1.8\times10^{-2}\ s^{-1}$	60	<7.7 %	21 s	39 s

4.3.2 Cycling Time and Amplification

The time required to complete one cycle is given by

$$t_{\text{one cycle}} = (k_1 + k_2)/k_1\, k_2. \tag{4-16}$$

In the example given by the first rows of Table 4-3, the cycling time is long compared to the time of approach to equilibrium and analyses using steady-state intensity measurements may be used without an appreciable degree of cycling. It is more a matter of reconstituting a component. In the last example with $k_2 = 0.2\ s^{-1}$ (bottom row), the rate constants are chosen such that a cycle is completed in one minute. Hence some amplification is caused by increased chemical conversion, which lets the reaction proceed through several cycles.

Compared with a two-step procedure cycling can be used to increase the amount of emitted light (i.e. the time integral of intensity vs. time):

$$A_t = \int_0^t I\,dt, \tag{4-17}$$

Integrating over a suffficently long time period allows the reaction to proceed through a number of cycles. The amplification may be kept at an arbitrarily high level by allowing a sufficiently long reaction time for the coupled analysis (Figure 4-5). The final limitation of the emitted amount of light is not determined by C, but by the availability of energy donor D. Amplification during cycles may also be understood in terms of a continued recharging of C during each cycle that the reaction is allowed to go through. The time allowed for the reaction determines the number of cycles and thus the amplification. This is different from conventional enzymatic cycling, where an accumulation takes place of a substance which is determined later. In this case, the amount of accumulated substance is proportional to the time allowed for the reaction. In luminescence analysis,

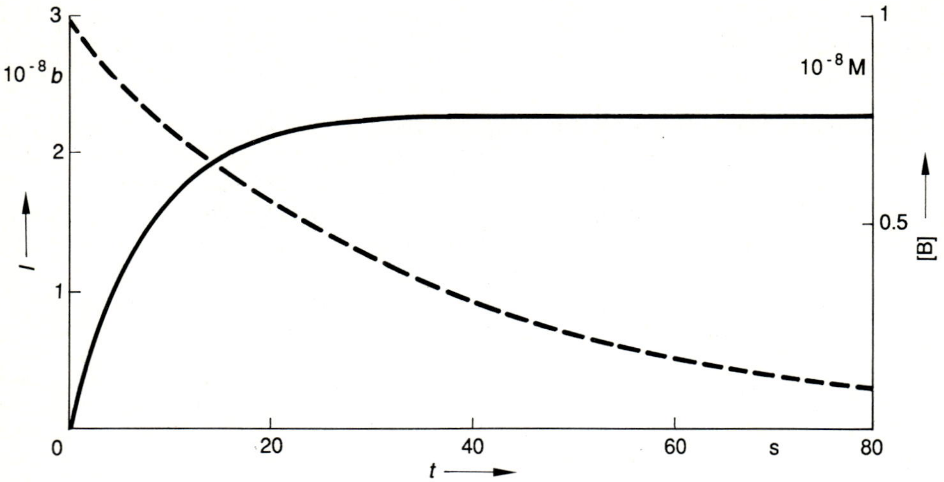

Figure 4-5. Analysis of a compound C through the reactions:

C + D → B rate constant k_1 (R 4-1)

(R +) B → light + C rate constant k_2 (R 4-2′)

The ordinate has two scales, one for the luminescence intensity *I*, and one for the concentration of B. It is assumed that equations (2-2) (2-3) apply, that $k_2 = 0.1\ s^{-1}$, and that the initial concentration of C is 10^{-6} M.

Dashed curve: Analysis in two separate steps. The luminescence reaction (R 4-2′) applied after a previous total conversion of C to B. Thus, it is assumed that a 10^{-6} M solution of C is converted to a 10^{-6} M solution of B. According to the simply theory, there is a light flash which decays exponentially, cf. equation (2-5).

Solid curve: Analysis with two reactions proceeding simultaneously, giving rise to cycles and a steady-state intensity, cf. equation (4-12). The curve is drawn with $k_1 = 0.03\ s^{-1}$.

the actual cycling rate is monitored. This approach allows the accumulation of photons in the recording equipment. The state of the equilibrium, i. e. the steady-state intensity, during cycling is lower than the peak height, which results, if $[C]_0$ is converted to $[B]_0$ in a separate preceding reaction, cf. Figure 4-5. This makes it necessary to apply light recording devices with a very stable background during measurement. This approach still makes it possible to receive more photons through integration over an extended time, including a large number of reaction cycles. Some experience with this type of problem has been obtained in radioactive analysis, where prolonged recording helps to distinguish a low signal from background noise.

4.3.3 Comparing Constant Light-Producing Methods

Cycling is particularly useful for the analysis of small amounts of nucleotides. The sensitivity improves considerably compared to a conventional two-step procedure with a preceding reaction which converts C to B. Except for high sensitivity applications, it is generally sufficient to measure the light intensity as soon as the steady-state is reached and thus to do without chemical amplification.

It is important to understand the difference between the constant light intensity produced by a coupled reaction cycle and that of a slowly reacting non-cycled luciferase system. In the former case, the light intensity is maintained by continued energy supply from a donor. In the latter case, it is a matter of extending the time profile of the emission by a very slow luminescence system, so that the emission appears to be essentially constant, but remains at a very low level. In other words, in most applications of slowly reacting luciferase systems, the nearly constant light intensities depend on slow fading and not on reconstitution by reaction cycles. It may be of interest to compare the light emission of a slowly reacting luciferase with that produced by reconstitution to obtain a truly constant signal. Assuming that slowly reacting luciferase has a fading rate of 1 % per minute, which is a normal requirement, the intensity would be roughly one half of that obtained with reconstitution under the conditions that lead to the lowest sensitivity in Table 4-3. Furthermore, it can be seen that reconstitution permits arrangements which cause a considerable increase in sensitivity while still permitting some loss of luciferase activity.

4.4 Kinetics of Coupled Reactions in Enzyme Analysis

4.4.1 General Application

In this paragraph, we shall discuss the kinetics for estimating the activity of an enzyme E, which catalyzes reaction (R 4-1), and is coupled to a luminescence reaction (R 4-2) in the same solution. The reaction scheme is again as follows:

$$C + D \xrightarrow{E} B \qquad \text{(R 4-1)}$$

$$(R +)\ B \xrightarrow{L} \text{light}\ (+\ C) \qquad \text{(R 4-2)}$$

Compounds C and D are kept in excess, and the rate of reaction (R 4-1) is solely determined by the activity of the enzyme. The reaction may thus be considered to be zero-order, the rate constant $k_{ES}[E]_0$ reflecting the activity of the enzyme. Reaction (R 4-2) is arranged such that it becomes first-order with respect to the light inducing compound B and may be assigned the rate constant k_2. In many cases, compound C is also regained in reaction (R 4-2), and the entire processes can be considered a cycle. However, as an excess of C is maintained, this does not affect the kinetics, albeit it helps to maintain an excess concentration of C. The overall kinetics are thus basically the same, irrespective of whether or not cycles are possible, i.e. whether or not C is produced in reaction (R 4-2). As before, the intensity of the emitted light can be taken as being proportional to the rate of reaction (R 4-2) as given by equation (4-2). The concentration of B is described by

$$[B] = k_{ES}\ [E]_0 k_2^{-1}\ (1 - \exp\{-k_2 t\}). \qquad (4\text{-}18)$$

There is thus a build-up of intensity, following the equation

$$I = b\, k_{ES}\, [E]_0\, (1 - \exp\{-k_2 t\}). \tag{4-19}$$

The enzyme activity may thus conveniently be determined either from the initial slope of the intensity curve or the steady-state intensity.

4.4.2 Determination from Initial Slope

In the intensity vs. time graph, the initial slope is given by the following term, derived from equation (4-19):

$$(\mathrm{d}I/\mathrm{d}t)_0 = b\, k_{ES}\, [E]_0\, k_2. \tag{4-20}$$

If b and k_2 are known, the enzyme activity is determined by the slope at zero time, Figure 4-6. This kind of analysis has only been carried out with a slowly reacting

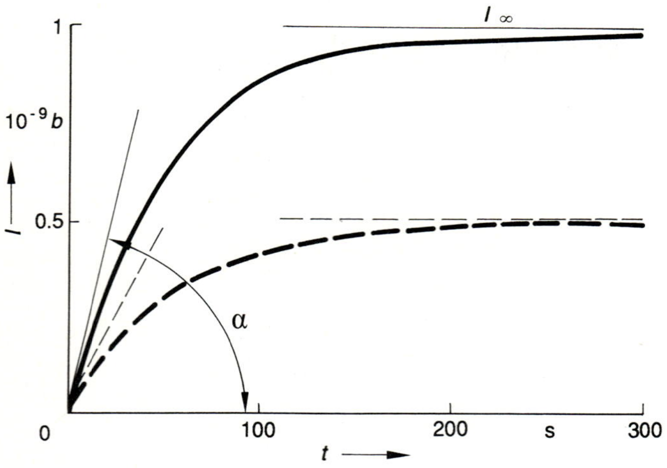

Figure 4-6. Illustration of how the activity, k_{ES} $[E]_0$, of an enzyme E can be determined by arranging a reaction scheme which includes a luminescence reaction. A coupled system of two consecutive reactions is applied:

$$C + D \xrightarrow{E} B \rightarrow \text{light}$$

The first reaction is made zero-order by supplying an excess of substrates and cofactors (C and D). By similar actions the luminescence reaction is made first-order in B. The rate constant of the luciferase reaction is fixed at $k_2 = 0.02\ \mathrm{s^{-1}}$.
Either the initial slope (angle α), or the constant light level (intensity I_∞) may be selected as analytical parameter; According to equations (4-20) and (4-21) tg $\alpha = (\mathrm{d}I/\mathrm{d}t)_0$ and I_∞ are both proportional to k_{ES} $[E]_0$.
Curves for two enzyme activities are shown:
$k_{ES}\ [E]_0 = 1.0 \times 10^{-9}\ \mathrm{M\ s^{-1}}$ (solid curve)
$0.5 \times 10^{-9}\ \mathrm{M\ s^{-1}}$ (dashed curve)
yielding $(\mathrm{d}I/\mathrm{d}t)_\infty = 2.0 \times 10^{-11}\, b$ and $1.0 \times 10^{-11}\, b\ \mathrm{s^{-1}}$ and $I_\infty = 1.0 \times 10^{-9}\, b$ and $0.5 \times 10^{-9}\, b$, respectively.

luciferase system [2, 9-11], although it is in principle also possible with a fast reacting luciferase. Equation (4-20) contains the rate constant k_2, and the method requires stable luciferase. The technique of stabilizing enzymes through immobilization is important in this context [4, 12, 13].

4.4.3 Determination From Steady-State

The second approach is based on measuring the steady-state intensity:

$$I_\infty = b\ k_{ES}\ [E]_0 \tag{4-21}$$

It should be noted that this intensity is independent of the rate of the luciferase reaction. A higher steady-state concentration of B compensates for a lower luciferase activity. The final intensity is proportional to the enzyme concentration $[E]_o$, Figure 4-6. With a fast reacting luciferase system, the intensity soon becomes constant, and the same intensity is ultimately attained also by a more slowly reacting system, cf. Figure 4-7. The

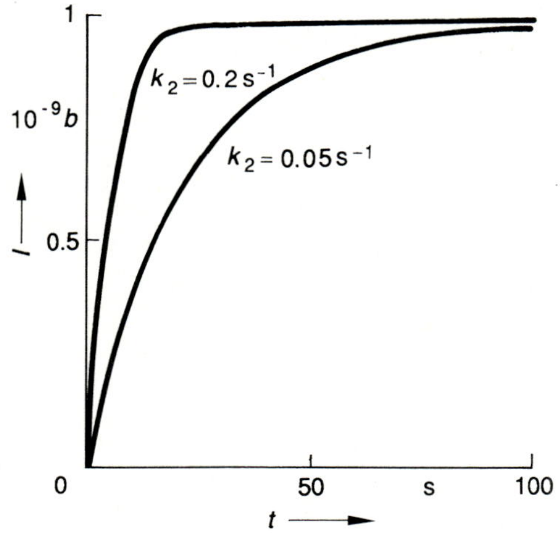

Figure 4-7. Determination of enzyme activity through coupling to a luminescence reaction. Effect of variations in luciferase activity is shown for a system in which the luciferase reaction is the much faster step. The system is set up as in Figure 4-6 with $k_{ES}\,[E]_0 = 1.0 \times 10^{-9}\,M\,s^{-1}$, but with $k_2 = 0.2$ and $0.05\ s^{-1}$, respectively. The same steady light intensity is rapidly approached and is reached independently of the condition of the luciferase, but it has to be ensured that the activity is suffiently high for convenient rapid analysis. The steady-state light intensity may thus be a better parameter for analytical determinations of enzyme activity than the initial slope, which carries with it some uncertainty about the stability of the luciferase.

insensitivity of the steady-state intensity to the luciferase activity may be of considerable analytical value in the assay of enzymes. The reaction time required to come sufficiently close to the steady-state is determined by the factor $\exp\{-k_2t\}$.

In Figure 4-8 experimental evidence is presented for kinetic compensation of decreased luciferase activity. The measured curves of intensity vs. time are in good agreement with theoretical curves, obtained by using equation (4-19) as in Figure 4-7.

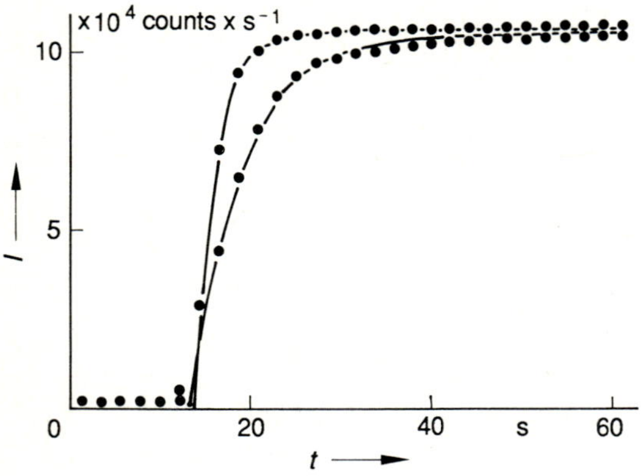

Figure 4-8. Coupled reactions designed for analysis of sorbitol dehydrogenase by determination of the steady-state level. The system is:

$$\text{NAD}^+ + \text{Sorbitol} \xrightarrow{\text{Sorbitol dehydrogenase}} \text{NADH}$$

$$\text{R} + \text{NADH} \xrightarrow{\text{Oxidoreductase/luciferase}} \text{light}$$

The dots give the experimental results in terms of photon counts collected during 2 s. The solid curves are simulations resulting from application of equation (4-19) using $k_2 = 1.0$ and $0.33\ s^{-1}$, respectively. Although the conversion rate was decreased as much as three times in the luciferase system (lower curve), the same steady-state intensity was obtained. From [14].

References

[1] Wettermark G, Brolin SE, Hjertén S. Cell Mol Biol. 1977; 22: 329–339.
[2] Lundin A., In: Kricka LJ. Carter TJN, eds. Clinical and Biochemical Luminescence. New York: Marcel Dekker; 1982: 43–74.
[3] Brolin SE, Naeser P, Wettermark, G. J Biolum Chemilum. 1989; 4: 446–453.
[4] Roda A, Girotti S, Ghini S, Carrea G. J Biolum Chemilum. 1989; 4: 423–435.
[5] Ågren A, Brolin SE, Hjertén S. Biochim Biophys Acta. 1977; 500: 103–108.
[6] Brolin SE, Ågren A, Wersäll JP, Hjertén S. In: Schram E, Stanley P, eds. Analytical Application of Bioluminescence. Westlake Village, California: State Printing & Publishing; 1979: 109–121.
[7] Brolin SE. Bioelectrochem Bioenerg. 1977; 4: 257–262.
[8] Stanley PE. In DeLuca MA, ed. Methods in Enzymology. New York: Academic Press; 1978: 181–188. (Bioluminescence and Chemiluminescence; vol LVII).
[9] Lundin A, Rickardsson A, Thore A. Anal Biochem. 1976; 75: 611–620.
[10] Lövgren T, Thore A, Lavi J, Styrelius I, Raunio R. J Appl Biochem. 1982; 4: 103–111.
[11] Raunio RP, Lövgren TNE, Kurkijärvi K. Anal Biochem. 1985; 150: 315–319.
[12] Jablonski E, DeLuca MA. In: DeLuca MA, ed. Methods in Enzymology. New York: Academic Press; 1978: 202–214. (Bioluminescence and Chemiluminescence; vol LVII).
[13] Ugarova NN, Brovko EY, Beliaieva EI. Enzyme Microb Technol. 1983; 5: 60–64.
[14] Brolin SE, Wettermark G. J Biochem Biophys Meth. 1991; 22: 1–18.

5 Experimental Design

5.1 Selecting Appropriate Analytical Systems

Although there exist a great many bioluminescence systems with potential analytical applications, only two of them have been used widely. These are the firefly and the bacterial luciferase systems. In both cases a number of typical reaction schemes have been worked out which are of practical importance and which may also be used in setting up other kinds of bioluminescence analyses. A brief discussion of other systems is included in the Chapter "Fields of Application".

In order to apply theoretical schemes, it is necessary to have some knowledge of the mechanism of native luminescence and to have access to data directly pertinent to the design of the analytical setup. This is a prerequisite for making the analyses both easy and reliable. However, of all available data, only a minor portion is needed for the design of a proper analytical setup. The desired information is often scattered among the general literature on bioluminescence. Hence, this chapter contains brief descriptions of the firefly and the bacterial luciferase systems, supplemented with collated information on matters such as conversion rates, optimal reaction conditions, K_m-values, and effects of inhibition. The presentation also contains characteristic examples of the application of theoretical schemes. Some fundamental aspects are described on how "a set of chemical reactions that produce light" may be developed into a useful analytical system. Experimental testing for applicability and reliability is directed to meet demands regarding sensitivy and convenient performance, particularly through the selection of appropriate rates for the various reaction steps. The examples are chosen to illustrate a proper and critical use of simple kinetic theory, as outlined in the preceding chapters.

5.2 The Firefly Luciferase System

5.2.1 Data Pertaining to the Planning of Assays

In an analytical procedure involving the firefly luciferase system, the substance to be analyzed (light inducer B, Chapter 2) is ATP. The luminescence reaction may be described by the scheme:

$$\text{D-Luciferin} + O_2 + \text{ATP} \xrightarrow[\text{Mg}^{2+}]{\text{Luciferase (L)}} \text{Oxyluciferin} + \text{PP}_i + \text{AMP} + CO_2 + \text{light} \qquad \text{(R 5-1)}$$

The reaction is thus an ATP dependent oxidative decarboxylation of D-luciferin which is accompanied by the emission of light (cf. ref. [1, 2]), with a very high quantum yield. A value as high as 0.88 has been reported [3]. The optimum pH is 7.75 [4], at which the emission peaks at 560 nm with a broad maximum [1, 3]. The luciferase reaction may be used over the entire range from pH 6 to pH 8, but the activity is pH-dependent. In analytical applications, it is thus important that the buffer capacity is high enough to ensure pH stability. At a low pH or in the presence of the cations Hg^{2+}, Zn^{2+}, or Cd^{2+}, there is a red shift of the emission spectrum [1]. The optimum temperature is between 23 and 26 °C, with a maximum at 25 °C [2, 5]. The structure of luciferase is known. It has a molecular weight of 61.000 [6] and contains sulfhydryl groups which are necessary for its catalytic activity [7, 8]. It has recently been produced through gene technology [6]. Firefly luciferase appears to be a very slow enzyme. Measurements with a purified preparation have indicated a value of 0.1 U mg^{-1} [9, 10].

In the design of analytical applications, it is advantageous to mentally divide the overall reaction into three distinguishable steps:

1. A rapid reversible reaction: Esterification of luciferin producing luciferyl adenylate and liberating inorganic pyrophosphate

$$LH_2\text{-COOH} + ATP + E \rightleftharpoons E\text{-}LH_2\text{-COO-AMP} + PP_i \quad \text{(R 5-2)}$$

2. A rate determining step [1], with a 25 ms time lag before light emission is initiated, and 300 ms before the peak intensity is reached: detachment of a proton and formation of a carbanion, which reacts with molecular oxygen to form excited oxyluciferin, which emits a photon.

$$E\text{-}LH_2\text{-COO-AMP} + O_2 \rightarrow E\text{-oxyL-AMP} + CO_2 + \text{light} \quad \text{(R 5-3)}$$

3. The slow dissociation of the enzyme product complex

$$E\text{-oxyl-AMP} \rightarrow E\text{-oxyL} + AMP \quad \text{(R 5-4)}$$

Step 1 is inhibited by the product pyrophosphate PP_i but under the prevailing conditions, this is normally ineffective. Step 2 defines the start-up process, which is usually only slightly affected by changes of the analytical conditions. Step 3 may lead to temporary occupation of the active site, if substrate concentrations are high.

Inhibition by pyrophosphate has long been known [11], but, on the other hand, low concentrations of pyrophosphate have also been reported to accelerate the turnover of luciferase [1, 11, 12]. Sulfhydryl reagents are very efficient inhibitors of the enzyme [7, 8]. Many other inhibitors are known including the reaction products AMP and oxyluciferin. They are regarded as competitive inhibitors with $K_u = 2.4 \times 10^{-4}$ and 2.3×10^{-7} M, respectively [13, 14]. However, it has been observed that oxyluciferin inhibition is much more effective, if the substance is produced in the luciferase reaction, than if it is merely added to the solution. This is believed to result from oxyluciferin remaining on the active site. Of great practical importance is the inhibition by analogs of D-luciferin, since addition of L-luciferin provides a basis for the development of the slowly reacting luciferase, available in commercial kits for "constant light level measurements" [9, 15]. The luciferase reaction is inhibited by salts, due to an increased K_m for MgATP at high ionic strength. In addition, there is a specific anion inhibition in the order $I^- > Br^- > Cl^- > F^- >$ carbonate > acetate [16]; $SCN^- > I^- \simeq NO_3^- > Br^- > Cl^-$ [17] and $ClO_4^- > I^- >$ acetate [18]. In this connection it may be mentioned that vanadate, which exerts a strong inhibitory effect on ATPase, does not have any disturbing inhibitory effect on the luciferase enzyme [19].

The generally accepted K_m value for ATP is on the order of 2.0-2.5 $\times 10^{-4}$ M [13, 20]. Thus, an acceptable range for ATP analysis may be below 1 μM. However, there are also indications that there exist two reactive sites for ATP binding, with K_m values of

1.1×10^{-4} and 2×10^{-5} M, respectively. Inhibition may concern one of the sites [21]. For luciferin, K_m is on the order of 2×10^{-6} M [22]. Taking into account the sequence of reactions which lead to light production, a mathematical expression has been worked out for evaluations of the emission time course [10].

5.2.2 Time Profile of the Light Pulse

According to the literature about the enzymatic characteristics of firefly luciferase, high substrate concentrations have generally been used. Primarily, the experiments have focussed on the clarification of reaction mechanisms rather than on models suitable for analytical applications. When it comes to analyzing ATP, an acceptable concentration range may be below 1 μM, which is much lower than the K_m, but yet high enough to allow for induction of a strong emission [2]. First-order kinetics should be provided for by the analytical setup, as discussed in Chapter 2. If, therefore,

$$-d[B]/dt = k_2 [B], \qquad (5\text{-}1)$$

where [B] is the concentration of the light inducer (in this case ATP) the rate constant k_2 should not be allowed to change during measurements. Theoretically, several causes for change of k_2 during the analyses may be considered. A change would occur if the active site of the luciferase were blocked by residual oxyluciferin. Other types of product inhibition may also be considered. Side reactions competing for ATP may, furthermore, disturb the measurements and give rise to apparent rate constants which do not represent the luciferase reaction.

Two approaches to a series of analyses are possible. In the first, the change of k_2 is small during the course of a recording and the reaction may be treated as first-order, yielding linear plots on a logarithmic scale. Changes of k_2 may occur between successive recordings and, for instance, represent a loss in activity, due to continuous degradation of the enzyme during the assay of several samples. In the second type, the emission data from a single recording are non-linear in logarithmic plots of intensity vs. time indicating that the overall reaction may not be represented as first-order. Such deviations may for instance be caused by inhibitors produced during the reaction or be the result of simultaneous parallel reactions.

The recording in Figure 5-1 may be considered to fall into the first category. Logarithmic plotting yields a straight line which is sufficiently linear, so that k_2 can be regarded as constant. The rate constant calculated from this diagram is 0.058 s^{-1}. In analyses employing this system, the light dose emitted during 50 s was recorded for the different samples. Here we are dealing with an integral representing the sum of counts recorded in the channels of a multichannel analyzer, together extending over 50 s. The selected time of integration is rather arbitrary but should include a major part of the total emission, without introducing large errors from the tailing end of the emission. With proper choice of time frame for this integrations, small variations of k_2 between the different runs may be accepted. The example, Figure 5-1, may be used to illustrate that the integral may not be very sensitive to small variations of k_2. Thus, using equation (2-7) it is found that a 5 % lowering of the rate constant only results in a 2 % reduction of the recorded amount of emitted light. If the rate constant goes down by 10 % the decrease in recorded counts will still be below 5 %. It is, of course, important

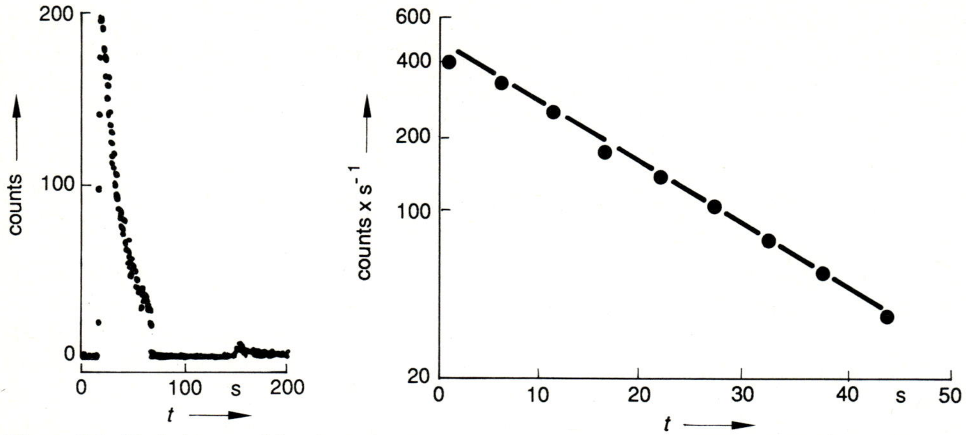

Figure 5-1. Emission resulting from 0.5 fmole ATP injected into a fast reacting firefly luciferase solution with purified enzyme. The small peak to the right represents a blank. The recording was carried out by photon counting and a multichannel analyzer [49]. Each dot represents the number of counts collected during 0.5 s. The integral covers 50 s resulting in the number of counts indicated above the peaks. A logarithmic representation of the sample peak is given to the right yielding a first-order rate constant of 0.058 s^{-1}.

to ensure that k_2 remains sufficiently stable during a series of measurements. In practice, the recording and evaluation of the entire flash profile is often troublesome and time-consuming. An easy test may be carried out by taking note of the ratio between the light emitted during an early phase of the flash as compared with that emitted during a later part, and to check that this ratio stays suffiently constant.

Figure 5-2 displays a recording of light flashes in which the emission cannot be represented by a single first-order rate constant. It is thus an example of a type of analysis which falls into the second category as outlined above. The initial part may follow first-order kinetics and be represented accordingly, but there is also a continuing longer-lasting emission. This later emission may be more or less pronounced, and its origin has been the subject of considerable debate. One hypothesis is that inhibitors are involved, cf. Figure 3-1. Another explanation may be that adenylate kinase (AK) is present and catalyzes the reaction

$$\text{AMP} + \text{ATP} \overset{\text{AK}}{\rightleftharpoons} 2\ \text{ADP} \qquad \text{(R 5-5)}$$

AMP may already be available in the original reaction solution, but it is also formed in the luciferase reaction. This implies that in the initial phase of the analysis, ADP may be formed through reaction (R 5-5) simultaneously with the luciferase reaction. Later on, when a large amount of ATP has been consumed for the production of light, reaction (R 5-5) will reverse and ADP will slowly recombine, forming ATP, which now feeds the luciferase reaction (R 5-1), giving rise to a long-lasting tail of the emission. Generally we have used reaction (R 2-2) to represent reaction (R 5-1) when discussing kinetical schemes of luminescence analysis.

Considerable amounts of ADP may occasionally be met with already at the start of the reaction. In some cases, special procedures have been undertaken to ensure that ADP is kept at a low level in the luciferase solution, in particular through the use of the apyrase reaction [23]. With large amounts of AMP present from the start, the slope of

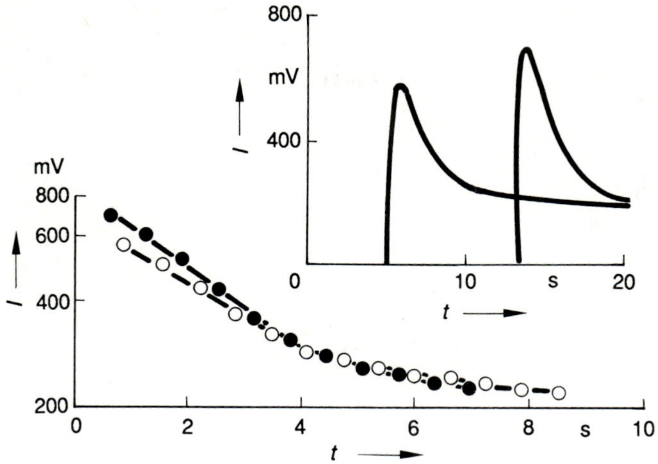

Figure 5-2. Emission profile obtained when analyzing ATP by using a crude extract of firefly luciferase.
Left curve: ATP extracted from a tissue sample of 110 ng dry wt.
Right curve: ATP standard of 1.8 pmole.
The data are represented logarithmically below. From the initial parts of the diagrams, apparent first-order rate constants can be calculated as 0.20 and 0.25 s^{-1}, respectively.

the initial emission may be determined by reaction (R 5-5) rather than by reaction (R 5-2). This is, in fact, the situation exemplified in Figure 5-2. In this case, the analysis was performed using a luciferase preparation which was known to be contaminated with AK and adenylate. The background emission elicited by the contaminants was essentially extinguished through treatment with apyrase, leaving AMP available for reaction with ATP as soon as the analysis was started. After a short rising time, an exponential decay took place and lasted for 4 s, as shown in the logarithmic plotting. The apparent rate constant for the disappearing ATP, as displayed by the immediate decrease in emission intensity after the peak and was calculated for a standard and a sample and found to be 0.25 and 0.20 s^{-1}, respectively. These high values indicate that it is a matter of apparent rate contants representing the overall consumption of ATP, both for the production of light and for the formation of ADP. Furthermore it can be seen that the later fading is no longer linear and finally becomes markedly retarded. Accordingly, we obtain a prolonged emission in agreement with the postulated supply of ATP to light formation, when the AK reaction is reversed. Figure 5-3 presents a data simulation of the light emission of ATP in a luciferase solution. It is illustrated how the presence of AK and AMP perturbs the exponential decay obtained with pure luciferase.

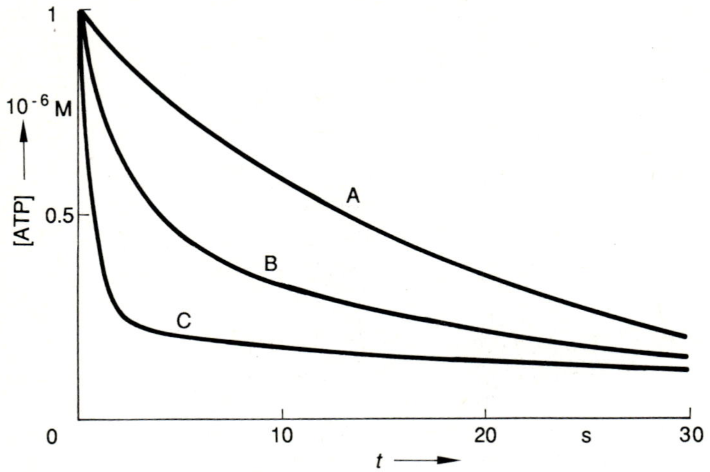

Figure 5-3. Calculated effects of the presence of adenylate kinase (AK) and adenosinemonophosphate (AMP) when analyzing 1.0×10^{-6} M ATP using the firefly luciferase reaction (R 5-1). It is assumed that:

ATP + AMP $\rightleftharpoons$ 2 ADP (R 5-5)

occurs parallel to the luciferase reaction, which is assumed to proceed with $k_1 = 0.05$ s^{-1}. Three cases are illustrated:

A: No AK present

B: 10^{-6} M AMP present at the start and sufficient AK present to let the forward reaction (R 5-5) proceed with a rate constant of 0.2×10^6 s^{-1} M^{-1}, yielding 0.088×10^6 s^{-1} M^{-1} for the reverse reaction, using the listed value of 2.26 [52] for the equilibrium constant of reaction (R 5-5).

C: Same conditions as B, expect for an AMP concentration of 5×10^{-6} M at the start of the reaction.

Although the initial exponential decay would recommend using the integrated emission intensity during this period as a measure of ATP, it should be realized that it is difficult to determine how much the measurements are disturbed through the delay between removal and supply of ATP. This lack of reliability may be illustrated by the difference between the rate constants, which causes a considerable fluctuation in the recording. In this situation, the peak height of the emission is preferred as a measure of ATP. During the short rising time, only a small fraction of ATP is converted to ADP; presumably a nearly proportional amount. It is a general experience that photokinetic techniques using measurement of peak heights have yielded accurate calibration curves.

Nowadays, sufficiently purified luciferase preparations are available, but the experience from analyses where reagents contained impurities is still of value when it comes to contaminations in the sample. To sum up: for fast reacting luciferase, it is generally safe to use the peak height as an analytical parameter. A small loss of ATP caused by side reactions during the rising time may be detected; and if significant, determined by calibrating with internal standards. On some occasions, integration over a fixed time period may be desirable in order to increase the sensitivity. In such cases, exponential fading should be ensured, as exemplified in Figure 5-1. The samples may, still indicate ATP concentrations that are too low, because of impurities which cause side reactions. This is revealed by apparently increased rate constants for the samples, as compared with those obtained with the standards, or alternatively by means of intercalated internal standards.

5.2.3 Nearly-Constant Light Emission

In analyses using a slowly reacting luciferase system, designed to yield a nearly-constant emission, the slow release of oxyluciferin from the enzyme is of minor importance, since there is ample time for reutilization and generally a comparatively high concentration of luciferase. The goal is to provide an analytical reagent which should permit degradation of no more than 1 % per min of a 1 μM solution of ATP [2]. In a first-order arrangement, this would correspond to a velocity expressed by $k_2 = 1.68 \times 10^{-4}\ s^{-1}$. As already mentioned, the rate retardation is mainly achieved by adding inhibitors, making a low conversion rate compatible with a high luciferase concentration. The reaction kinetics and the requirements for an analytical setup have been presented in detail [9, 15].

The analyses may be disturbed in the same way as described for fast reacting luciferase but other techniques are employed for detecting irregularities. In particular, deviations from nearly-constant intensity have been found to be useful as an immediate sign of disturbances. Alterations of the rate constant k_2 are directly expressed through change of emission intensity. In the case of gradually impaired luciferase activity, the intensity displayed will decrease. Removal of ATP through side reactions will correspondingly result in a decline; whereas deceptive supply is revealed by an intensity increase. Other alterations of k_2 which do not give rise to a downward or upward slope can be detected by a changed intensity level. The presence of inhibitors in a sample may, for instance, decrease the rate of light formation. Adding ATP to such a sample reveals such errors, if the emission increase is compared with a standard. It is often possible to replace routine standards by adding ATP to the samples. Such internal standards compensate for a long-lasting change of light production, but may fail to do so in case of interference through side reactions. Calculations therefore become uncertain, if interfering reactions consume or produce ATP at different rates before and after the addition of internal standards. Hence, photokinetic evaluations may be required particularly for the detection of velocity changes, as expressed by gradually increasing or diminishing intensities during the emission. However, it is generally accepted that the technique of using a slowly reacting luciferase system producing nearly-constant emission is often sufficiently reliable for analyzing ATP in a series of samples using only a few separate standards and blanks.

5.2.4 Assay of Phosphotransferases

The slowly proceeding firefly luciferase reaction may be regarded as a continuous indication of the ATP concentration, cf. Section 4.2.2 for $k_2 << k_1$. It can therefore be used conveniently to follow changes in the ATP concentration as an effect of reactions in which ATP is involved. Procedures for substrate analysis have been worked out, cf. Section 5.2.5, but the advantage of using ATP analysis as an indicator is particularly obvious in the assay of enzymes involved in the production of ATP, e.g. various phosphotransferases such as pyruvate kinase and creatine kinase [2].

A determination of adenylate kinase (AK) according to reaction (R 5-5) is shown in Figure 5-4. The amount of adenylate kinase added has to be low enough to make the

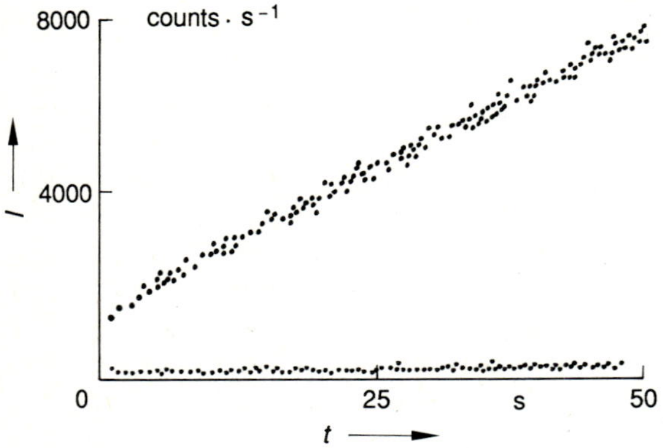

Figure 5-4. Design of coupling to firefly luciferase for the analysis of adenylate kinase activity. The emission was recorded from 100 μL slowly fading luciferase solution containing 5 μM ADP, to which was added a 5 μL sample of 1.8 μU adenylate kinase to initiate the reaction. Each dot represents the emission during 0.5 s. A linear intensity increase is observed. The rate of increase is proportional to the enzyme activity.
The signals were displayed on the screen of a multichannel analyzer, sample above, blank below.

consumption of ADP negligible during the measurement. According to the specifications given by the enzyme supplier (Boehringer Mannheim GmbH), the enzyme activity corresponds to $k_{ES}\,[E]_0 = 17$ mU $L^{-1} = 300$ pM s^{-1}. Although it is possible to saturate with the substrate (ADP) and to work at consequently high levels of ATP formation, it is often better to lower the rate and thus the ATP concentration by selecting an ADP concentration below the K_m value (K_m(ADP) $= 1.58 \times 10^{-3}$ M). In Figure 5-4, a photon counting technique was employed and the levels of ADP are about three orders of magnitude below K_m. The rate of ATP formation stays constant, yielding a nearly linear emission increase. The intensity increase becomes proportional to the enzyme activity, but it should be emphasized that the method requires calibration with an enzyme of known activity. A practical method for this kind of analysis has been worked out for use with a photo multiplier and a tracking voltmeter allowing the ATP concentrations to reach much higher levels than those given in Figure 5-4 [24].

The above method exemplifies measurements based on the formation of ATP. It is also possible to use the consumption of ATP to evaluate enzyme activities. Due to the low concentrations of ATP in luminescence analysis, it is often necessary to carry out analyses of an enzyme far below its K_m for ATP. Such low concentration causes a decreasing substrate level during the analysis. Accordingly, the kinetics are more complicated and calibrations should preferably be based on known quantities of the enzyme to be assayed.

5.2.5 Substrate Analysis

The scheme in Figure 5-5 illustrates a frequently used analytical procedure, which is based on direct coupling to the firefly luciferase reaction [2, 4]. Compounds which can

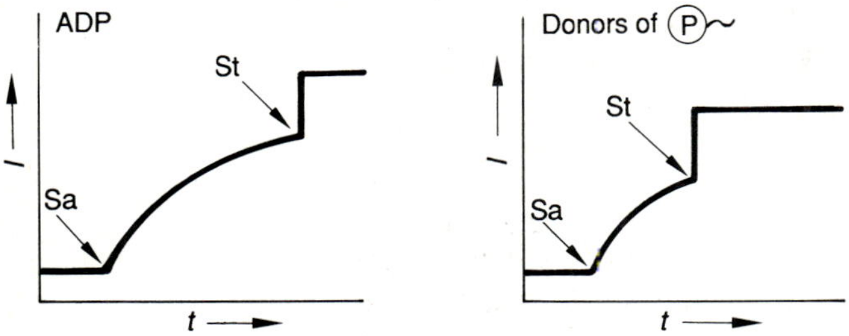

Figure 5-5. Determination of ADP (left) and phosphate donors (right) using a reaction directly coupled to a slowly reacting firefly luciferase system. Addition of sample (Sa) results in relaxation at higher intensity. When this level is sufficiently closely approached, an ATP standard (St) is added. The two intensity increments are compared to estimate the amount of ATP formed. ADP and PEP (phosphoenolpyruvate) may, for instance, be analyzed conveniently using the pyruvate kinase reaction.

participate in reactions in which ATP is produced or consumed can be assayed. The addition of the sample results in gradual relaxation to a new intensity level. Usually the substrate to be analyzed is added to a luciferase solution which already contains the necessary additives. The change in intensity level is compared with the result of adding an ATP standard. The system should be set up in such a way that it results in a sufficiently fast relaxation, so that it is essentially completed within the time allowed for measuring a sample. During this time, the consumption of ATP for light production in the luciferase reaction should be negligible; the analyses being set up with a luciferase system which fades at a rate of under 1 % per minute. Applications based on ATP consumption may sometimes be awkward because of the long relaxation times, due to the low ATP concentration allowed in the luciferase reaction.

As pointed out above for enzyme analyses, the luciferase reaction serves as an indicator of the ATP concentration at any given time. The procedure allows safe comparisons between samples of different kinds and is therefore a useful tool for the convenient evaluation of experiments. Furthermore, in most cases the analyses give a satisfactory approximation of the true quantities. Most preparations of firefly luciferase for durable emission are stable enough to make the frequent use of internal ATP standards unnecessary, if series of samples are analyzed.

Analyses of ADP or PEP (phosphoenolpyruvate) using the pyruvate kinase reaction are examples of applications involving ATP formation. The reaction is:

$$\text{PEP} + \text{ADP} \xrightarrow{\text{pyruvate kinase}} \text{ATP} + \text{pyruvate} \qquad \text{(R 5-6)}$$

In ADP analyses, surplus amounts of PEP are added; in PEP analyses, there should be an excess of ADP. Instead of employing standards with ATP, it may sometimes be better to prepare standards of ADP and PEP, respectively, and allow them to be processed together with the samples and the blanks. This may be preferable, as a complete conversion to ATP – although often closely approached – is never possible within the time frame allowed for the reaction. A few additional ATP standards may ensure unimpaired luciferase activity.

5.3 The Bacterial Luciferase System

5.3.1 The Two Enzymes

Among the different types of luminescent bacteria, there are three groups of marine bacteria whose extracts have been employed in luminescence analysis. There is some ambiguity in the nomenclature, and the table below may serve as a guide to reading the literature [25].

Group A: *Vibrio fischeri* = *Photobacterium fischeri* = *Achromobacter fischeri;*
Group B: *Photobacterium phosphoreum; Photobacterium leiognathi;*
Group C: *Beneckea harveyi* = *Lucibacterium harveyi* = *Photobacterium fischeri* MAV = *Vibrio harveyi; Beneckea splendida; Vibrio cholerae;*

The light-producing reaction of different species of luminescent bacteria is essentially the same (cf. [26, 27]). It can be considered a two-step reaction, catalyzed by an oxidoreductase (OR) and a luciferase (L):

$$NAD(P)H + FMN + H^+ \xrightarrow{OR} FMNH_2 + NAD(P)^+ \qquad \text{(R 5-7a)}$$

$$FMNH_2 + R\text{-}CHO + O_2 \xrightarrow{L} \text{light} + FMN + R\text{-}COOH \qquad \text{(R 5-7b)}$$

In the first reaction, NADH or NADPH reduces FMN with the help of an oxidoreductase, producing $FMNH_2$. In the second step, $FMNH_2$ reacts with a long-chain aldehyde with the aid of bacterial luciferase and light is emitted. This occurs through a long sequence of reactions [26, 28, 29], and the final step is thought to consist of an energy transfer (from an excited FMN species) to a protein-bound fluorescent agent named lumazine [30]. The emission is predominantly in the blue-green spectral region with maximum close to 500 nm. Although lower than for firefly luciferase, the quantum yield is very high compared to non-biological luminescence. A value as high as 0.27 has been reported [31], cf. Section 1.3.3.

The difference between the various groups of luminescent bacteria concerns both the oxidoreductase and the luciferase moieties. In *Photobacterium fischeri* the reduction of FMN by NADH and NADPH is catalyzed by oxidoreductase, with a K_m value of about 5×10^{-5} M for NADH [32]. From *Beneckea,* two varieties of oxidoreductase have been separated, one specific for NADH and the other for NADPH [33]. Both have K_m values in the area of 4-5 $\times 10^{-5}$ M. In other investigations, a slightly lower value for the NADH specific enzyme was reported [34]. The activities of the two enzymes differ slightly, amounting to $k_{ES}[E]_0 = 30$ µmol min^{-1} mg^{-1} for NADH and 50 µmol min^{-1} mg^{-1} for NADPH [33]. A 90 % signal reduction is observed when NADPH is used as substrate for the NADH-specific oxidoreductase, whereas the reduction of the signal is as high as

99 %, if, instead, NADH is tested against the NADPH specific enzyme [33]. The temperature maximum lies higher and is more distinct for the NADPH- than for the NADH-specific enzyme [35]. The pH profiles of the oxidoreductases do not differ greatly. Luciferases extracted from different species may have different pH maxima, but in the overall reaction thus including oxidoreductase and luciferase, an optimum at pH $\approx$ 7 has been reported for several enzyme preparations [35-37].

Regarding luciferase, the quantitative response to aldehydes differs between enzymes with the species. All react with various long-chain aldehydes, but sensitivity to chain length varies. *Photobacterium fischeri* displays an intensity about ten times higher for dodecanal (12 C) than for decanal (10 C) [28]; while the response of *Beneckea harveyi* is essentially the same for all aldehydes from octanal (8 C) to dodecanal [36]. It may also be of analytical importance that preparations from *Beneckea*, in contrast to those from *Photobacterium*, can be considerably inhibited by an excess of aldehydes [36]. All reported K_m values for $FMNH_2$ and aldehydes are low. Studies of luciferase from *Photobacterium fischeri* show that K_m values for $FMNH_2$ are at or just below 4×10^{-5} M [38, 39]. For aldehydes, K_m values are given as $K_m = 4.7 \times 10^{-5}$ M for *Photobacterium fischeri* – decanal; and $K_m = 2.7 \times 10^{-6}$ resp 6.5×10^{-5} M for *Beneckea harveyi* – octanal and dodecanal [40]. The specific activity of the luciferase is commonly given with respect to the number of emitted photons (and not to the amount of reacted substrate). Among luminescent bacteria, there are mutants yielding so called dark pathways, i.e. the same type of overall scheme as reaction (R 5-7), but without light emission [29]. The dark pathways appear not to interfere with conventional analyses.

Luciferase is rapidly inactivated by imidazole (0.1M) irrespective of the pH; whereas oxidoreductase is essentially unaffected [39]. Although mainly of interest in connection with studies of reaction kinetics, it may be mentioned that either the luciferase site for $FMNH_2$ or that for aldehyde binding can be blocked separately by means of different inhibitors [40]. A feature common to the oxidoreductases and the luciferases is the abundance and sensitivity of SH groups, which are liable to be affected by thiol reagents [41, 42]. It is a well-established fact that several enzymes are inactivated by dodecylsulfate, which likewise applies to the bacterial luciferase system. It has long been known that it is also inhibited by some salts [43]. Inhibition by cations e.g. Zn^{2+} and Ca^{2+} has also been reported [44]. Bacterial luciferase is very susceptible to the action of protease, which may even quench the emission [45, 46]. The presence of hydrazine derivatives may decrease the light signal due to the capture of aldehydes.

Both $FMNH_2$ and aldehydes bind firmly to their specific sites. The attachment of $FMNH_2$ occurs quickly and stays unaffected, until the production of light starts. Bound $FMNH_2$ is in this way protected from a direct nonenzymatic oxidation; whereas unbound $FMNH_2$ is oxidized by dissolved oxygen in the reaction mixture ($t_{1/2}$ = 0.1 s [47]). This occurs faster than the enzyme sites become accessible again after completion of the light reaction. Consequently, it is necessary to design an analytical system in such a way that any $FMNH_2$ is immediately taken care of by the luciferase, as soon as it is formed.

In analyses of NAD(P)H, $FMNH_2$, or aldehydes, the K_m values given in the preceding sections may provide some guidance in chosing appropriate concentrations. The concentration of the compound to be determined should be considerably below its K_m value and concentrations of other reactants markedly above their respective K_m's.

Hence, it is often suitable to keep the concentration of the target compound below μM concentration by diluting the sample, which also diminishes disturbances due to side reactions. The K_m values of other reactants indicate that it is generally easy to obtain sufficiently high concentrations. In fact, there may be an inclination to overrate the need for high concentrations.

5.3.2 Emission Profiles of NADH and NADPH

As discussed in Chapter 2 and explained above in connection with analyses of ATP with firefly luciferase, an appropriate analytical setup requires first-order kinetics. With reference to equation (5-1), the rate constant k_2 should be essentially unaltered as long as the signal is used for measurements. This applies to both fast and slowly reacting bacterial luciferase systems.

Figure 5-6 exemplifies recordings during an NADH analysis with a fast reacting system. Rapid injection of NADH results in a flash which then fades essentially

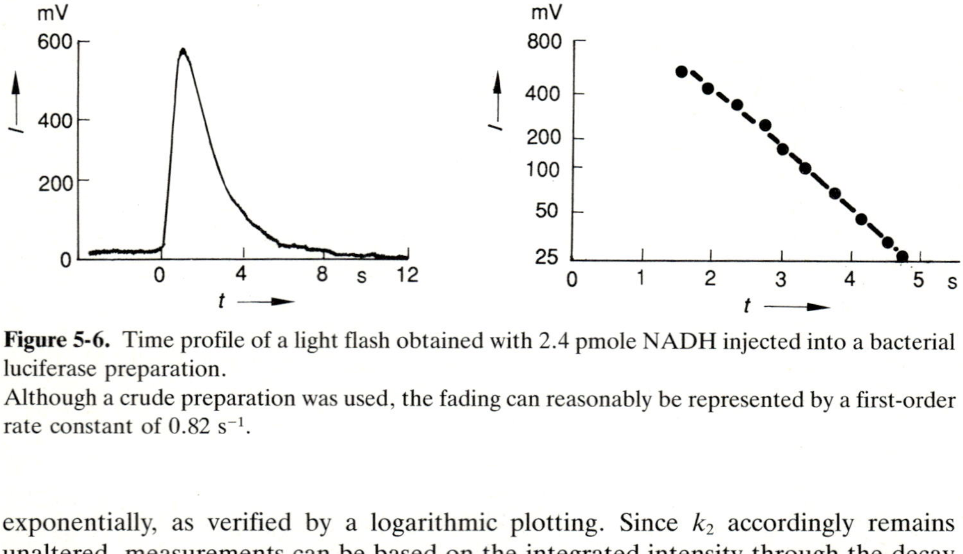

Figure 5-6. Time profile of a light flash obtained with 2.4 pmole NADH injected into a bacterial luciferase preparation.
Although a crude preparation was used, the fading can reasonably be represented by a first-order rate constant of 0.82 s^{-1}.

exponentially, as verified by a logarithmic plotting. Since k_2 accordingly remains unaltered, measurements can be based on the integrated intensity through the decay period. However, it is advisable to integrate over a fixed time, in order to exclude disturbances by possible variations in the long-term decay of the emission. If k_2 varies, alternative use of the maximum as analytical parameter should be considered. The maximum registered in a photokinetic recording may stay unaffected, despite the appearance of disturbing side effects during the later part of the emission. This approach is well-established and sustained by the linearity of calibration curves. The relation between a high value of the rate constant k_2 and high emission intensity should be appreciated, and may provide the basis for enhancing analytical sensitivity.

The convenience of a more durable emission intensity than a flash was demonstrated in connection with the use of the 3-acetylpyridine analogue of NADH for the reduction of FMN [37]. In contrast to the peaks produced by NADH the signals with the analogue stayed essentially unchanged for many seconds, facilitating the readouts. Arrangements have been developed and are commercially available as kits for the analysis of NAD(P)H by monitoring a nearly-constant emission. This provides for a convenient readout, as illustrated in Figure 5-7. Such analyses with slowly reacting bacterial

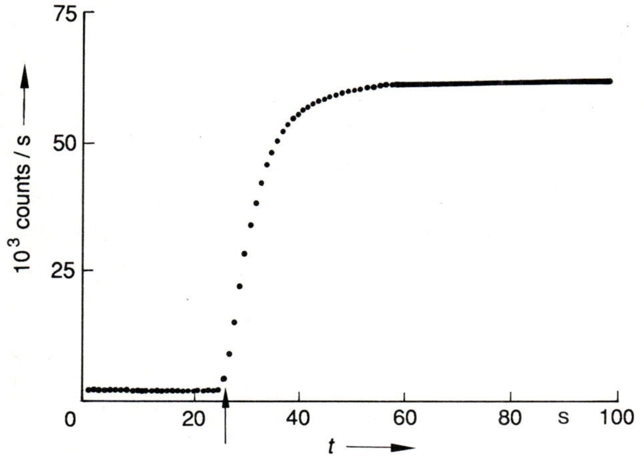

Figure 5-7. Emission time course for the slowly reacting bacterial luciferase system. A 10 μL solution containing 1 nmol NADH was rapidly injected (arrow) into 200 μL light producing solution from a commercial reagent kit (LKB). The solution contained phosphate buffer with unspecified amounts of oxidoreductase and luciferase, isolated from *Beneckea Harveyi,* about 1.1 μM FMN, and a surplus of decanal. The concentration of FMN is below the K_m of *Beneckea* luciferase [34] and thus contributes to the prolonged production of light, which is mainly a result of low oxidoreductase concentration. In this test, NADH reaches a very high concentration, 4.5 μM, which, however, is below the K_m for oxidoreductase, but not in favor of a long-lasting nearly-constant emission.
Light recording principle:
The recording was obtained using a photon counting procedure and a multichannel scalor to store the information [49]. The height of each dot represents the number of counts collected during a preset time, here 1 s, which are displayed chronologically. For convenient reading, a "number of count" scale and a time scale have been added to the photograph.

luciferase require that the measurements and calibrations are checked much more carefully and frequently than required for the corresponding assays of ATP with firefly luciferase. The lower reliability seems to depend on a fundamental difference in the design of the two systems. When employing the firefly system, the enzyme (luciferase) is supplied at fairly high concentrations, and the rate is limited by added inhibitors. With the bacterial luciferase system, the conversion rate is primarily regulated by the oxidoreductase concentration, which is kept low. Inactivation of a small quantity of enzyme, for instance through surface denaturation, will accordingly only have a small effect on the light output of the firefly system but may result in a large decrease of the light intensity in analyses involving the bacterial luciferase system. It might be helpful to use immobilized or encapsulated oxidoreductase, which protects the enzyme from inactivation.

A useful alternative for obtaining a stable emission is the enzymatic reconstitution of reduced pyridine nucleotides as described in Section 4.3. In this design, the emission is kept at a more "truly" constant intensity by means other than ensuring a sufficiently low conversion rate through a reduction of the oxidoreductase concentration. Moreover, both the oxidized and the reduced nucleotides can be measured without mutual interference, because enzymes can be chosen for the specific reduction of either NAD^+ or $NADP^+$ (cf. Figure 7-1).

5.3.3 Applying the Kinetics of Coupled Reactions

Although the bacterial luciferase reaction can in many instances be represented by a single overall rate constant, it should be kept in mind that it represents a coupled reaction system composed of two simultaneously ongoing reactions, i. e. the oxidoreductase and the luciferase reactions. As mentioned above, this coupling determines the variations in the time course of the emission. Figure 5-8 shows how varying the amount

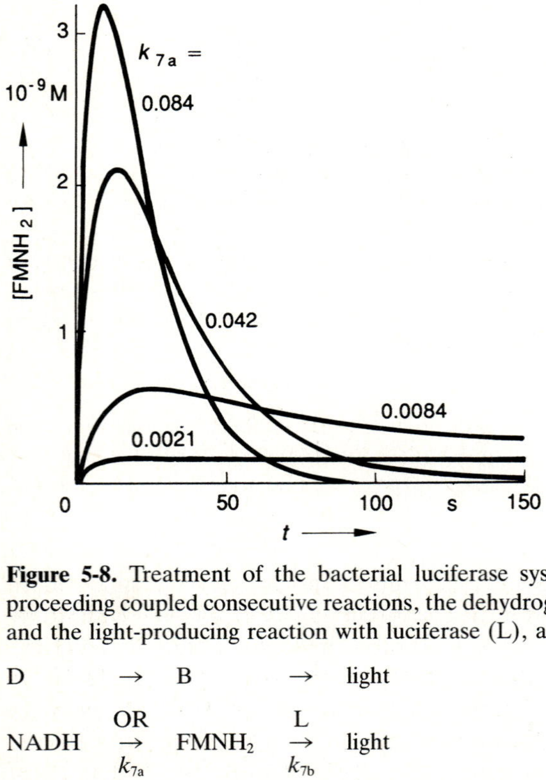

Figure 5-8. Treatment of the bacterial luciferase system as a system of two simultaneously proceeding coupled consecutive reactions, the dehydrogenase reaction with oxidoreductase (OR) and the light-producing reaction with luciferase (L), assuming $[NADH]_0 = 1.0 \times 10^{-8}$ M.

$$D \rightarrow B \rightarrow \text{light}$$

$$NADH \underset{k_{7a}}{\overset{OR}{\rightarrow}} FMNH_2 \underset{k_{7b}}{\overset{L}{\rightarrow}} \text{light} \qquad \text{(R 5-7)}$$

The emission profile is simulated using equation (4-3) by varying the relation between the activity of OR and L, here expressed in terms of the first-order rate constants k_{7a} and k_{7b}. The intensity is then proportional to the concentration of the intermediate species $FMNH_2$. The curves assume the same luciferase activity, $k_{7b} = 0.11\ s^{-1}$, and for a varying oxidoreductase activity. It is shown how the emission may be changed from that of a flash to a more continuous emission by varying the activity relation between the two enzymes.

of oxidoreductase in relation to luciferase may change the emission from an initial flash to a later maximum, or to a long-lasting nearly-constant emission. Experimental data have demonstrated how the delay can be extended, resulting in a slowly fading and hence practically nearly-constant emission [34]. In reality, there are limited possibilities of regulating the steps leading from the oxidation of NAD(P)H to the production of light. Thus, as pointed out earlier, the luciferase activity must be high enough to immediately bind any $FMNH_2$ as it is formed, preventing oxidation by oxygen dissolved in the reaction solution. The requirements for luciferase depend on the rate of the FMN reduction, which in turn is determined by the activity of oxidoreductase and the concentrations of NAD(P)H and FMN. In analytical practice, it may be safe to work with an eight- to tenfold weight excess of luciferase [27]; but this is not always necessary, since the requirements are determined by the rate of the enzymatic FMN reduction.

The specific activity of bacterial luciferase is declared to be much lower than that reported for oxidoreductase, which points to practical difficulties concerning the supply of sufficient amounts of luciferase for binding and keeping pace with the rapid production of $FMNH_2$. It must be kept in mind, however, that statements about enzyme activities generally refer to situations where enzyme concentrations are low compared to substrate concentrations, and where enzymes are therefore reutilized. The time required for an enzyme molecule to become accessible for a new turn depends not only on the time for the substrate to bind to the specific site and on the time required for the actual reaction, but also on the relaxation time, i. e. basically, the time required for the reaction products to leave the site. For luciferase, it is believed that the relaxation time dominates, which means a higher activity, if the enzyme is not part of a cycle. A non-cycled process often arises in luminescence analysis, when the system is saturated with luciferase and is designed to yield a fast light flash.

Figure 5-9 represents an experiment in which an attempt was made to compose a reaction mixture yielding the emission maximum at a time convenient for appropriate mixing and recording. Testing was carried out with light-producing solutions, as described in the caption. The enzymes were separated from *Photobacterium fischeri.* Two curves are given. The luciferase concentration is the same in both cases, whereas the oxidoreductase is 4.5 times lower in the second display. Simulating the emission by using equation (4-3) ($k_1 = k_{7a}$ and $k_2 = k_{7b}$) yields the rate constants $k_{7a} = 0.0125\ s^{-1}$ and $k_{7b} = 0.05\ s^{-1}$. It may be interesting to try to estimate the specific activities of the enzymes from these rate constants. Obviously, only very rough estimates may be obtained. A straightforward calculation, based on $K_m = 5 \times 10^{-5}$ M for NADH/oxidoreductase and $K_m = 4 \times 10^{-5}$ M for FMNH/luciferase, yields the activities 37 and 120 U L^{-1}, respectively, in the reaction mixture, and the specific activities 18 U mg^{-1} for oxidoreductase and 3 U mg^{-1} for luciferase. The first value is on the same order of magnitude as a reported specific enzyme activity of about 100 U mg^{-1} enzyme. For luciferase, a direct comparison is more difficult, as questions of reaction efficiency (quantum yield) enter into any comparison of reported activities.

The necessity of carrying out a proper calibration under the prevailing conditions during analysis should be pointed out. There is usually a linear relationship between maximum height and quantity of the compound to be determined. In order to minimize disturbances from the mixing of samples and reagents when evaluating the kinetics of the reaction, the recordings shown in Figure 5-9 were obtained by using an injection

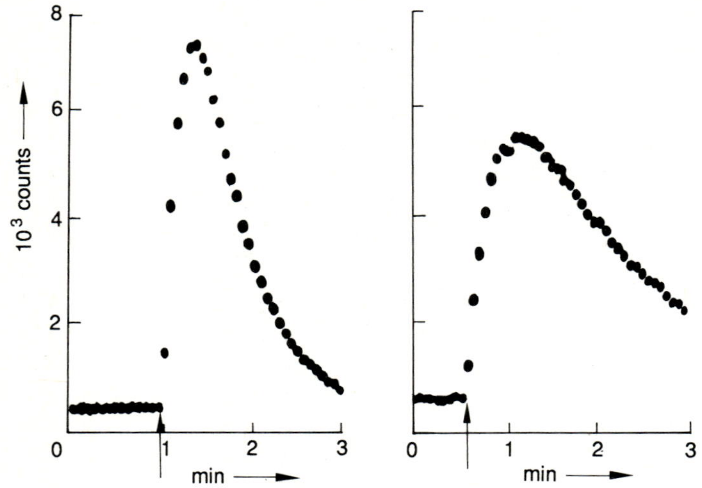

Figure 5-9. Designing the bacterial luciferase system for convenient determination of maximum intensity, cf. [53]. The recording procedure is the same as used in Figure 5-7 but with a preset time of 4 s for each intensity reading.
Manual mixing and positioning of the reaction vessel call for a sufficient delay in the appearance of the maximum. By employing an injection technique, it is possible to determine the time available for such measures. Samples of 10 μL, containing 10 pmole NADH, were rapidly injected (arrowed) into 240 μL of a light-producing solution. The two displays show the effect of lowering the oxidoreductase concentration by a factor of 4.5, moving the position of the maximum from 24 to 36 s after starting the reaction.
Composition of the Reaction Solution:
At the start of the reaction, the light-producing solution is composed of 0.08 M sodium phosphate buffer at pH 6.8, containing 0.04 % BPA, 1.8 M glycerol, 0.25 mM EDTA, 77 μM dithiothreitol, 26 μM FMN, 20 μM myristic aldehyde, and 38 μg mL^{-1} luciferase. The oxidoreductase concentrations are 7.6 and 1.7 μg mL^{-1}, respectively. The purified enzymes are isolated from *Photobacterium fischeri* (Boehringer Mannheim GmbH), and the amounts given refer to the weight of enzyme protein. On the molar basis, the amount of luciferase would match the formation of about 45 pmol $FMNH_2$ without reutilization.

technique. In analytical practice, manual mixing and positioning of samples is often preferable. A simple and inexpensive recording technique using a tracking voltmeter is often sufficient to reach the required sensitivity, which is maintained at a sufficiently high level to permit dilution of samples, minimizing the role of unwanted side reactions. The stability of the solutions is a crucial factor which has to be checked. An initial increase of the activity of the reagent solution is frequently observed. In a procedure in which reagent solutions are prepared first and then left for an hour at room temperature before starting the measurements, the stability is usually sufficient to allow analysis during the following hour without the need for frequent calibration.

5.3.4 Substrate Analyses Through Direct Coupling

The analyses discussed above refer basically to situations in which the species to be measured, i.e. the nucleotide or the aldehyde, has previously been extracted as a test sample. Obviously, the same analytical situation exists if, instead, the light-inducing species has been produced in a preceding reaction designed for the analysis of substrates or enzymes. This particularly concerns assays based on the formation of NAD(P)H in a dehydrogenase reaction. In this case, it is seldom necessary to let this preceding reaction continue to completion, but a preferable approach is to allow time for sufficient accumulation of the species. The high sensitivity of luminescence often permits very short reaction times and, accordingly, only a small degree of conversion, but enough for an accurate analysis. However, prolonged incubation may be used to allow increased accumulation when there is a need for enhanced sensitivity.

Instead of prior accumulation of a light-inducing compound it may be possible to link the two reactions directly and to carry them out simultaneously in a common reaction mixture, cf. Chapter 4. In principle, we encounter the same kinds of disturbances in this type of scheme, as when the reactions are carried out as separate steps. Consequently, it is advisable to check the reactions by carrying out the individual reactions separately beforehand. It is also well-advised to test for any deviations from the standards by adding known amounts of nucleotide to the test samples and comparing the emission increase with the calibration curve.

An illustration of the use of direct coupling between a dehydrogenase reaction and the bacterial luciferase system for substrate analysis is given below. Some important steps when making such a system analytically suitable are presented. The goal was to obtain an emission time curve with an easily recognizable intensity maximum for quantifying the results. The delay in the appearance of the maximum should be in line with manual mixing of sample and reagents. The example concerns the analysis of sorbitol involving the sorbitol dehydrogenase reaction. The reaction was started by rapid addition of sorbitol with an injection technique, permitting a more precise evaluation of the kinetics.

In the scheme of coupled reactions, the dehydrogenase reaction is thus:

$$\text{Sorbitol} + \text{NAD}^+ \rightleftharpoons \text{Fructose} + \text{NADH} \qquad \text{(R 5-8)}$$

while the light-producing reaction is represented by the overall bacterial luciferase system, reactions (R 5-7). The set of coupled reactions (R 5-8) and (R 5-7) in this chapter may thus be identified as reactions (R 4-1) and (R 4-2). Reaction (R 5-8) above is reversible but may be easily shifted to the product side by an excess of NAD^+. In order to obtain the highest sensitivity in the set of coupled reactions (cf. Chapter 4) we attempt to adjust the two reactions (R 5-8) and (R 5-7) to proceed with rate constants on the same order of magnitude, in this case, about $10^{-1}\ s^{-1}$ for convenient recording. From the emission obtained with NADH added to the bacterial luciferase system, as shown in Figure 5-9, we know that k_7 fulfills this requirement at the given enzyme levels. The K_m value of sorbitol is reported to be in the range of $0.1 - 1 \times 10^{-3}$ M [48]. A value of 5×10^{-4} M would lead to a required sorbitol dehydrogenase activity of about 30 U L^{-1}, in order to obtain $k_a \simeq 10^{-1}\ s^{-1}$. Figure 5-10 shows the results of an initial set of experiments with sample composition, as reported in the caption. A spontaneous emission increase is observed, obviously due to some side reaction. Sorbitol addition gives rise to a rapid increase of the emission level, but a maximum cannot be discerned as the emission continues to increase. Nevertheless, it should be possible to analyze sorbitol with the above composition of reagents, using the increased emission level as a basis for the calibration curve. However, such analyses would probably give rise to

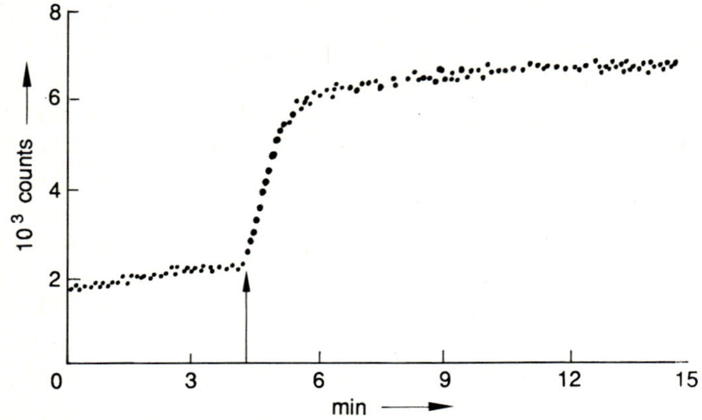

Figure 5-10. Developing an analysis of sorbitol based on a coupled set of reactions with sorbitoldehydrogenase and the bacterial luciferase system.
Stage 1, initial approach:
A 10 μL sample containing 10 nmol sorbitol was injected (arrow) into 250 μL of a solution with the same basic composition as described in Figure 5-9 (with the lower oxidoreductase composition), but also containing 1mM NAD^+ and 0.65 μg mL^{-1} (26 mU mL^{-1}) sorbitoldehydrogenase. Recordings as in Figure 5-7. The emission sampling period was 4 s.
There is an initial strongly increasing background emission at the level of 500 counts s^{-1}. Injection of sorbitol causes a marked rise of the emission, which is somewhat difficult to quantify accurately, because of the ongoing spontaneous emission increase.

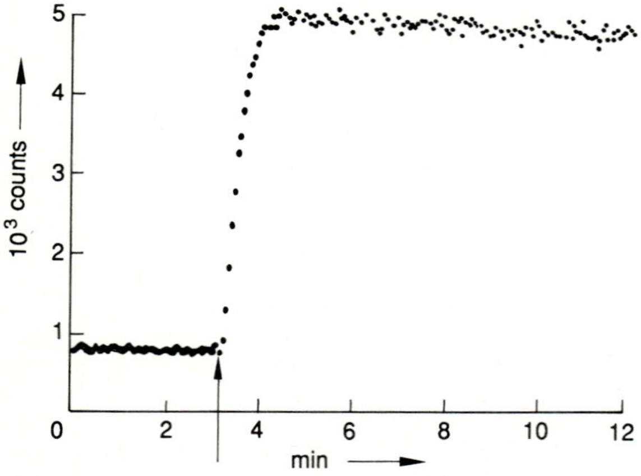

Figure 5-11. Developing an analysis of sorbitol based on a coupled set of reactions with sorbitoldehydrogenase and the bacterial luciferase system.
Stage 2, improvements through limiting reactions that are responsible for background emission:
A sample of 10 nmol sorbitol in 10 μL was injected (arrow) into 250 μL of a solution with the same basic composition as described in Figure 5-10, except for an increased sorbitoldehydrogenase concentration of 11.5 μg mL^{-1} (460 U L^{-1}) and without glycerol. Recordings as in Figure 5-7. The emission sampling period was 4 s.
The background emission is steady. Addition of sorbitol causes a rapid increase of the emission and is followed by a gradual decrease of the emission. Hence, measurements at maximum intensity are possible.

analytical difficulties of various kinds. Further development of the method is thus required and would include trying to eliminate the side reaction responsible for the increasing background. One would also try to increase the sorbitol dehydrogenase activity in order to provide a maximum. The side reaction proved to be the oxidation of glycerol, which had been added for protection of the oxidoreductase enzyme. Figure 5-11 shows the results obtained after removal of glycerol and with a ten times higher sorbitol dehydrogenase activity.

5.3.5 Assay of Enzymes with Direct Coupling

Direct coupling between a preceding enzyme reaction and the bacterial luciferase system may be employed for the determination of enzyme activity [50, 51]. An example is shown in Figure 5-12. As described in Chapter 4, it is a matter of determining a

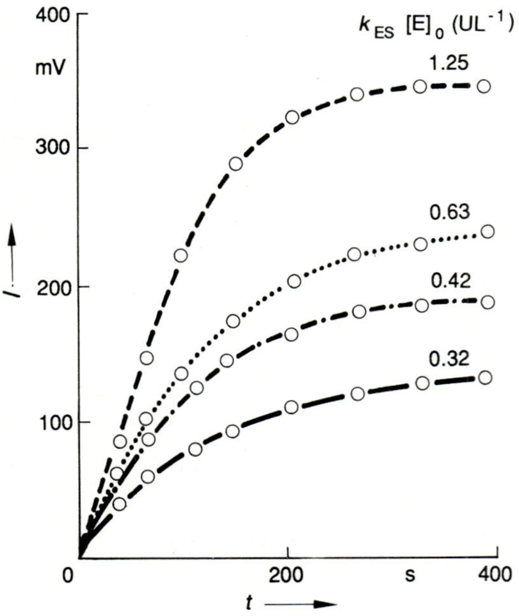

Figure 5-12. Investigation of aldose reductase using coupling to bacterial luciferase via aldehyde formation, cf. reactions (R 5-9) and (R 5-10). Emission in mV, plotted versus reaction time. The activity of aldose reductase is indicated on each curve as μU present in the test solution. A 115 μL portion of a luciferase assay mixture (Boehringer Mannheim GmbH), but in the absence of myristin, was supplemented with 1.7 mM NAD^+, 0.9 μM NADH, and 5 μL myristyl alcohol dissolved in ether (150 mg mL^{-1}). It was allowed to fade for about 10 min, in order to extinguish the light from contaminating aldehyde. The reaction was started by adding a 5 μL sample of aldose reductase with the activity indicated on the graph.

steady-state intensity as opposed to a maximum, as in the corresponding substrate analyses in the preceding paragraphs.

The two coupled reactions may be described as:

$$NAD^+ + \text{myristyl alcohol} \xrightarrow{\text{aldose reductase}} NADH + \text{myristic aldehyde} \quad \text{(R 5-9)}$$

$$\text{myristic aldehyde} + FMNH_2 + O_2 \xrightarrow{\text{bacterial luciferase}} \text{myristic acid} + FMN + H_2O + \text{light} \quad \text{(R 5-10)}$$

An excess concentration of $FMNH_2$ is maintained through the presence of oxidoreductase and NADH in the reaction solution. Equations (4-18) – (4-21) should be applicable for the overall event. As indicated in Chapter 4, this kind of analysis has the advantage of yielding a response in terms of a steady emission, independent of variations in luciferase activity, because the latter only affects the time needed to approach the steady-state.

The approach to the steady-state is determined by the luciferase reaction and may be approximated with the first-order rate constant k_{10}. Equation (5-2) yields a logarithmic representation of Equation (4-19) suitable for plotting the data:

$$\ln (I_\infty - I) = \ln I_\infty - k_{10}\, t. \quad (5\text{-}2)$$

It is seen from Figure 5-13, left graph, that all four plots may approximately be represented by the same rate constant $k_{10} \simeq 1.0 \times 10^{-2}\,s^{-1}$. The graph to the right shows

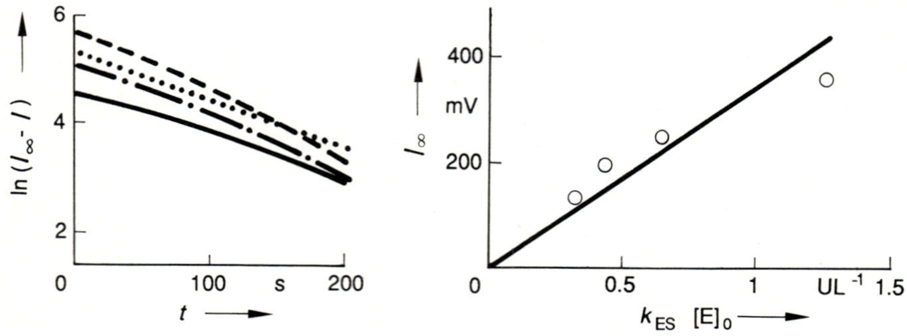

Figure 5-13. Graphical treatment of the data of Figure 5-12, using the theory given in Chapter 4. The approach to the equilibrium is determined by the bacterial luciferase reaction $k_{10} \simeq 1.0 \times 10^{-2}$ s^{-1}. The steady-state intensity I_∞ is then proportional to the aldose reductase activity.

that equation (4-21) holds and may be used to determine the constant b, yielding $b \simeq 350\ mVLU^{-1}$

In essence it is shown that the emission intensity at the steady-state (I_∞) may serve as a convenient measure of enzyme activity.

References

[1] DeLuca M, McElroy WD, In: DeLuca MA, ed. Methods in Enzymology New York: Academic Press; 1978: 3–15. (Bioluminescence and Chemiluminescence, vol LVII).

[2] Lundin A, In Kricka LJ, Carter TJN, ed. Clinical and Biochemical Luminescence. New York: Marcel Dekker; 1982: 43–74.

[3] Seliger HH, McElroy WD. Arch Biochem Biophys. 1960; 88: 136–145.

[4] Lundin A, Richardson A, Thore A. Anal Biochem. 1976; 75: 611–620.

[5] McElroy WD, Strehler BL. Arch Biochem Biophys. 1949; 22: 420–433.

[6] Hill F, Wood K, DeLuca M. In: Schölmerich J, Andreesen R, Kapp A, Ernst M, Woods WG, eds. Bioluminescence and Chemiluminescence New Perspectives. Chichester: John Wiley & Sons; 1987: 397–400.

[7] DeLuca M, Wirtz GW, McElroy WD. Biochem. 1964; 3: 935–939.

[8] Lee R, McElroy WD. Biochem. 1969; 8: 130–136.

[9] Wulff K, Haar H-P, Michal G. In: Serio M, Pazzagli M, eds. Luminescent Assays Perspectives in Endocrinology and Clinical Chemistry. New York: Raven Press; 1982: 47–52.

[10] Schram E, Ahmad M, Moreels E. In: DeLuca MA, McElroy WD, eds. Bioluminescence and Chemiluminescence Basic Chemistry and Analytical Applications. New York: Academic Press; 1981: 491–496.

[11] McElroy WD, Hastings WD, Coulombre J, Sonnenfeld V. Arch Biochem Biophys. 1953; 46: 399–416.

[12] Ahmad M, Schram E. In: DeLuca MA, McElroy WD, eds. Bioluminescence and Chemiluminescence Basic Chemistry and Analytical Applications. New York: Academic Press; 1981: 435–441.

[13] Lee TR, Denburg JL, McElroy WD. Arch Biochem Biophys. 1970; 141: 38–52.

[14] Goto T, Kubota J, Suzuki N, Kishi Y, In: Cormier MJ, Hercules DM, Lee J, eds. Chemiluminescence and Bioluminescence. New York: Plenum; 1973, 325–331.

[15] Lundin A, In: Serio M, Pazzagli M, eds. Luminescent Assays Perspectives in Endocrinology and Clinical Chemistry. New York: Raven Press; 1982: 29–45.

[16] Aledort LM, Weed RI, Troup SB. Anal Biochem 1966; 17: 268–277.

[17] Denburg JL, McElroy WD. Arch Biochem Biophys. 1970; 141: 668–675.

[18] Gilles R, Requeux JJ, Saive JJ, Thome-Lenz G. Arch Int Phys Biochim. 1976; 84: 807–818.

[19] Baltus E, Hanocq-Quertier J, Schram E. In: Schölmerich J, Andreesen R, Kapp A, Ernst M, Woods WG, eds. Bioluminescence and Chemiluminescence New Perspectives. Chichester: John Wiley & Sons; 1987: 567–570.

[20] Lee YS, Crispen RG. In: Borun G, ed. Proceedings of the 2nd Biannual ATP Methodology Symposium. San Diego: SAI Technology Co; 1977: 219–235.

[21] DeLuca M, McElroy WD. J Biochem Biophys Res Com. 1984; 123: 764–770.

[22] Hatefi Y, Hahstein WG. Proc Nat Acad Sci. 1969; 62: 1129–1136.

[23] Lyman GE, DeVincenzo JP. Anal Biochem 1967; 21: 435–443.

[24] Brolin SE, In: Bergmeyer HU, ed. Methods of Enzymatic Analysis, vol III. Weinheim: Verlag Chemie; 1983: 540–559.

[25] Hastings JW, Nealson KH. Ann Rev Microbiol. 1977; 31: 549–595.

[26] Hastings JW. In: DeLuca MA, ed. Methods in Enzymology New York: Academic Press; 1978: 125–135 (Bioluminescence and Chemiluminescence, Vol. LVII).

[27] Jablonski EG, DeLuca M. In: Kricka LJ, Carter TJN, eds. Clinical and Biochemical Luminescence. New York: Marcel Dekker; 1982: 75–87.

[28] Hastings JW, Spudich J, Malnic G. J Biol. Chem. 1963; 238: 3100–3105.

[29] Hastings JW, Balny C. J Biol Chem. 1975; 250: 7288–7293.

[30] Small ED, Koka J, Lee J. J Biol Chem. 1980; 255: 8804–8810.

[31] Hastings JW, Riley WH, Massa J. J Biol Chem 1965; 240: 1473–1481.

[32] White CL. High Sensitivity Analytical Determination of Nicotineamide Adenine Dinucleotide and Coupled Enzyme Systems with the Bacterial Bioluminescence Reaction [Dissertation] Athens, Georgia: University of Georgia; 1980.

[33] Jablonski E, DeLuca M. Biochem. 1977; 16: 2932–2936.

[34] Lavi J, Raunio R, Malkov Y, Lövgren T. Biochem Biophys Res Com. 1983; 111: 266–273.

[35] Lövgren T, Peacock R, Lavi J, Karp M, Raunio M. International Laboratory. 1982; July/August.
[36] Hastings JW, Weber K, Friedland J, Eberhard A, Mitchell GW, Gunsalus A. Biochem. 1969; 8: 4681–4689.
[37] Brolin SE, Borglund E, Tegnér L, Wettermark G. Anal Biochem. 1971; 42: 124–135.
[38] Puget K, Michelson AM. Biochimie 1972; 54: 1197–1204.
[39] Brolin SE, Hjertén S. Mol Cell Biochem. 1977; 17: 61–73.
[40] Nealson KH, Hastings JW. J Biol Chem. 1972; 247: 888–894.
[41] Friedland J, Hastings JW. Biochemistry 1967; 6: 2893–2900.
[42] Nicoli M, Meighen EA, Hastings JW. J Biol Chem 1974; 249: 2385–2392.
[43] McElroy WD, Hastings JW, Sonnenfeld V, Coulombre J. J Bacteriol. 1954; 67: 402–408.
[44] Stanley PE. Anal Biochem. 1971; 39: 441–453.
[45] Baldwin TO. In: DeLuca MA, ed. Methods in Enzymology, New York: Academic Press; 1978: 198–201 (Bioluminescence and Chemiluminescence, Vol. LVII).
[46] Baldwin TO, Hastings JW, Riley PL. J Biol Chem. 1978; 253: 5551–5554.
[47] Meighen EA, Hastings JW. J Biol Chem. 1971; 246: 7666–7674.
[48] Leissing N, McGuiness ET. Biochem Biophys Acta 1978; 524: 254–261.
[49] Wettermark G, Stymne H, Brolin SE, Petersson B. Anal Biochem. 1975; 63: 293–307.
[50] Brolin SE. Mol Cell Biochem. 1983; 55: 177–182.
[51] Brolin SE, Berggren PO, Naeser P. In: Schölmerich J, Andreesen R, Kapp A, Ernst M, Woods WG, eds. Bioluminescence and Chemiluminescence New Perspectives. Chichester: John Wiley & Sons; 1987: 519–522.
[52] Eggleston HV, Hems R. Biochem J. 1952; 52: 156–160.
[53] Brolin SE, Wettermark G. J Biochem Biophys Methods. 1991; 22: 1–18.

6 Preparation of Samples

6.1 Requirements

Analyzing a pure sample of a compound is frequently vastly different from applying the same kind of analysis to a “real situation”, such as determining the composition of a biological sample. This is also the case in bioluminescence determinations. Thus, a great variety of bioluminescence methods have been published, including details of kinetics, model experiments, and calibration curves. They can, in general, be applied directly to samples in which disturbances have been eliminated by preceding purification steps, for instance, the effluent from chromatographic columns. To be applicable in cell research, further development of the analytical procedure is usually required. The preparation of the sample is particularly important. In fact, accuracy and reproducibility in biochemical analyses depend sometimes more on the preparation of the specimens than on the final chemical measurements. This has long been known in the application of absorption spectroscopy or fluorescence, but it is also true for bioluminescence analysis.

In cell rersearch, it is important to be aware of concentration changes in the cell, which take place during the initial phase of sample preparation. The goal may be the measurement of a transcient metabolite, calling for immediate prevention of conversion. Losses of the compound to be determined must likewise be minimized during further processing of the specimen. In addition, interfering substances must be controlled and their effects sufficiently eliminated. To satisfy these demands, various preparative techniques have been developed. The total preparative procedure includes several steps, with options for further treatment, procedure control, and quantification measures. It is advantageous to use diagrams to gain an overview over these matters. Although the preparation and analysis of single samples may sometimes proceed in continuous and rapid succession, a break should generally be planned. In this way, a series of samples can be collected, and after quantification processed ready for analysis. The measures to be taken in order to maintain the metabolic state naturally depend on the length of the intermission, which may be limited to the time required for the collection of the samples, or extended to their storage.

Before entering into a detailed discussion of the various ways of collecting and processing the samples, attention must be paid to some fundamental measures involved in the preparative procedures. These measures concern arresting the metabolic processes, lyophilisation, storage of specimens, adjustment of concentrations, and modes of quantification. Together, they prepare samples ready for extraction and analysis. Details of suitable extraction procedures are given later in this chapter.

6.2 Preserving Species Subject to Assay

6.2.1 Arresting the Metabolic Processes

Freezing is the predominant technique for interrupting metabolic conversions. In principle, freezing enables safe preservation, but this is [illegible] true when it comes to practical applications. The method chosen for the initial quick-freezing must be regarded in relation to the conversion rate of the compound to be measured. It deserves to be pointed out that, due to the high thermal inertia of water, 10 to 20 s are needed to freeze an organ to a depth of 10 mm, if its surface is brought to liquid nitrogen temperature [1]. Fast-freezing of a tissue can be accomplished by cutting off part of an organ of an anaesthetized animal and dropping the piece into a beaker with isopentane or freon chilled by liquid nitrogen to −150 °C, Figure 6-1. Dipping it directly into liquid nitrogen is not useful, since a protective layer of evaporated gas immediately covers the specimen and retards the chilling, unless the nitrogen temperature is lowered below its boiling point by partial evacuation. When organs are difficult to move to the body surface without interrupting the blood circulation, metal tongs chilled in liquid nitrogen may be used and applied so that a thin frozen specimen is formed, Figure 6-2. Blood that is partially squeezed out at the edges of the tong is removed by a chilled scraper. An interruption of the oxygen supply may quickly change existing equilibria and thus alter the concentration of adenine and pyridine nucleotide forms (cf. Figures 8-2, 9-10, 9-11). The demands for powerful freezing may be reduced in small specimens, which can be quickly frozen because of their high surface-to-mass ratio. The handling of liquid nitrogen may therefore be replaced by chilling isopentane to −70 °C by dry ice, as in experiments with pancreatic islets weighing less than 5 µg.

Cultured cells or cells in fluids may be frozen as pellets after centrifuging them briefly in their medium, provided that substances that interfere with the analysis do not need to

Figure 6-1. Preserving metabolites by fast-freezing.
The metabolism is rapidly arrested when an organ from an anaesthetized animal is brought to the body surface with an uninterrupted vascular circulation and then plunged into isopentane chilled to −160 °C.

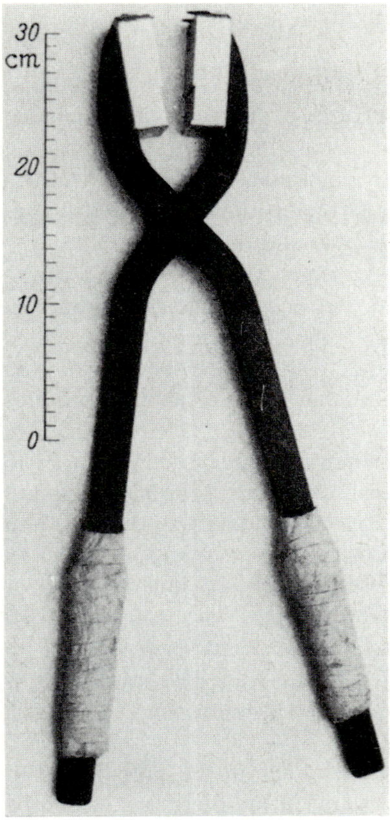

Figure 6-2. The original appearance of metal tongs designed for freeze-clamping.
The tongs are chilled to the temperature of liquid nitrogen. Metal pieces with a high thermal conductivity contribute to quick freezing. In principle, no significant improvements of the technique have hitherto been envisaged. From [1].

be washed out. Body fluids may also be frozen to conserve the metabolic processes, but certain precautions have to be taken to prevent efflux from suspended cells, e.g. hemolysis of erythrocytes. Removal of cells by repeated centrifuging is a standard procedure.

Although freezing is widely applicable for arresting metabolic processes, other means may be preferable or sometimes needed for the preparation of samples for bioluminescence measurements. Most enzymes are not damaged by rapid freezing and can be stored and assayed in the frozen samples, together with the metabolites. However, a minority of enzymes are affected by a freezing process but can be measured immediately after homogenization in a cold hypotonic buffer. Relevant activity changes are scarcely encountered during the first 30 min in an ice bath. It is generally not recommended to analyze metabolites using such homogenates, but total amounts of nucleotides can be determined, as they are not affected by interconversion among the respective forms. For simplicity in performance, it is sometimes desirable to be able to prepare samples without using cryotechniques. This can be achieved by extraction that is accompanied by simultaneously arresting the metabolism [2,3].

6.2.2 Lyophilization and Cryostat Techniques

Removal of water from frozen tissues makes them less inclined to undergo biochemical alterations. The technique is generally called lyophilization or freeze-drying. Weighing at room temperature is possible, and storage conditions are improved. Furthermore, capillary forces within the lyophilized specimens provide rapid contact with extraction fluids. Immediate suppression of undesired reactions is possible through pH adjustment and supplementation of the extraction fluid with suitable additives.

Frozen specimens, such as pellets of cultured cells or small cultured organs and also small pieces of organs, can be directly lyophilized, but it is important to prevent temperature increase during transfer to the evacuation chamber. The use of a chilled metal case with sufficient thermal capacity is recommended. Organ specimens often contain groups of different cells. Isolating cells of the same kind for analysis requires special techniques (cf. Figure 8-7).

The cryostat sectioning technique is of great value when applying bioluminescence to the microanalysis of tissues. A cryostat for histological sectioning consists of a cabinet equipped for freezing and a microtome operated from outside. It is the standard equipment for sectioning frozen specimens for histochemical staining. Applied to quantitative microanalysis, it is recommended to cut sections of 5–10 μm at −25 °C, or 20–25 μm at −30 °C. It is convenient to provide the cryostat with accessories for subsequent lyophilization. The freeze-drying may be carried out at −30 °C in an evacuation tube. This tube is connected to the vacuum pump via a cold trap, which is immersed in a Dewar filled with dry ice / heptane. Evacuation for more than 8 hours at a pressure below 1.3 Pa (0.01 mmHg) is required.

The drying of the frozen specimen is often a critical stage in the preparation of samples for analysis. A small amount of condensed water inside the evacuation tube may jeopardize the process. It is important for the equipment to have a sufficient drying capacity. A simple device of the kind described above may be excellent for cryostat sections, but cannot be used with larger specimens.

6.2.3 Checking Alterations During Intermissions and Storage

In principle, freezing arrests metabolic processes and provides for safe preservation. Accordingly, storage and handling in a cold state is the preferred means of sample conservation during the preparation of samples for analysis. However, serious disturbances may be caused by recrystallization in frozen undried tissue. The transformation between solid and liquid phase brings dissolved metabolites into contact with enzymes. It is thus a slow degradation, which nevertheless may be fast enough to have disastrous consequences during storage. Enzymes may be damaged by oxidation of SH-groups by dissolved oxygen in the liquid phase. Recrystallization may also bring about profound changes in their structural conformation.

Below −90 °C, the eutectic point of ice and water, no recrystallization takes place, allowing excellent storage conditions. However, a low temperature box with a nominal storage temperature of −70 °C is usually quite satisfactory for storage times not exceeding a few weeks. It is important that the actual storage temperature is held below

−50 °C, otherwise enzyme activity may give rise to changes. However, temperatures between −35 °C and −40 °C might allow a few days of storage. The rapid deterioration in a sample at higher temperatures may be illustrated by the observation that a 20 % decrease in ATP concentration occurs after 24 hours at −25 °C, after 4 hours at −15 °C, and already after 30 min at −10 °C, as reported from the storage of frozen mouse brain [4]. Considerable changes were also observed with other metabolites. Most rapid decrease showed glucose-6-phosphate with a loss of 20 % after one hour of storage at −20 °C, or 10 min at −10 °C.

The removal of water in a specimen improves storage conditions. This is one basic reason for using lyophilized samples. Such samples also have the advantage of being easily deoxygenized. At −20 °C in a vacuum or under an argon atmosphere, they may be stored for weeks without any detectable biochemical change.

6.2.4 Adjustment of Concentrations

In biochemical analysis, it is often possible to reduce the effect of interfering reactions caused by contaminants in the sample by simply diluting the solution to be analyzed. This can be explained as follows. In principle, biochemical reactions are kinetically dependent on the concentration of an enzyme and on one or several substrates. Upon dilution, the rates of such reactions are reduced at least quadratically, provided that the concentrations are sufficiently below the K_m values. A desired analytical reaction can be arranged such that its rate is only reduced linearly upon dilution. The desired concentration of all species, i.e., enzymes and substrates necessary for the reaction, can be maintained through the addition of reactants; except, of course, the species to be analyzed. By extensive dilution, it is often possible to practically eliminate the effect of other reactions, but this obviously requires a highly sensitive analytical method. In many cases, bioluminescence meets this requirement.

Adjustments of concentrations following the principles outlined above are illustrated in Figure 6-3 for the case of an analytical reaction

$$C + D \xrightarrow{E} B \qquad \text{(R 6-1)}$$

and an interfering reaction

$$C + D' \xrightarrow{E'} B\ . \qquad \text{(R 6-2)}$$

C is a cofactor supplied to a give a high concentration in the reaction mixture. D is the substrate for the enzyme E and D' for the enzyme E'. The substrate D or the enzyme E is the target compound, determined through measurements of B (using bioluminescence). In the analysis of D, the concentration of E is maintained at an appropriately high level. In the assay of E, there is an excess of D in the reagent, i.e. sufficient to give a concentration in the reaction mixture well above the K_m value. Dilution affects the concentration of the species to be assayed linearly – and, accordingly, the rate. In the interfering reactions, both D' and E' are diluted, and thus the effect on the rate will be quadratic.

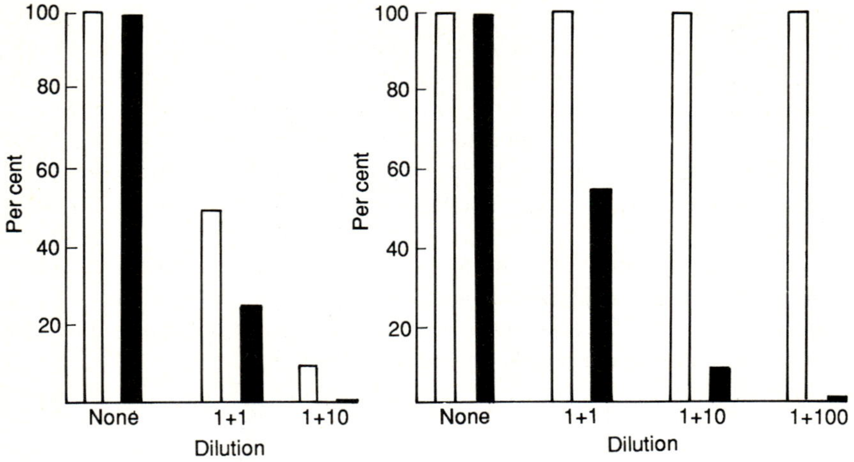

Figure 6-3. Illustration of the advantage of diluting a sample to provide a better ratio between an analytically used wanted reaction (open bars) and an unwanted reaction (filled bars).
The reactions are assumed to be catalyzed by an enzyme and to take place between a reagent and an excess of cofactor. Dilution decreases the rate of the unwanted reaction quadratically, due to decreased concentrations of both the enzyme and its substrate. In the reaction employed for analysis (wanted reaction), only the substrate (or enzyme) to be measured is diluted. Hence the rate decrease is linear.
Left: Absolute reaction rates compared – For both reactions the rate before dilution is given the value 100 %.
Right: Relative reaction rates compared – Rate of the analytical reaction (wanted reaction) and the unwanted reaction in relation to that of the wanted reaction.

Low concentration resulting from extensive dilution frequently calls for protective measures against denaturation. Despite the small volumes involved in microtechniques, the amount of a sample may be so small that very low concentrations are encountered from the start. Many compounds and enzymes are liable to undergo decomposition, particularly at the walls of tubes or other containers. Surface denaturation affects many enzymes at protein concentrations below 0.02 %, but can usually be counteracted by the addition of bovine plasma albumin. Below μL level, the high surface-to-volume ratio calls for the protein concentration to be raised above 0.02 %, but between 0.05 and 0.1 % should suffice.

6.3 Quantification of Samples

The cell itself is a unit and thus an excellent quantification measure, either for a single cell assay or for cell counts in culture experiments. Very frequently, the amount of a component is expressed in terms of the weight of the sample. However, handling frozen specimens may entail difficulties. Lyophilization allows easy and safe weighing, since removal of the water prevents disturbing reactions long enough to permit handling the samples for an hour or two at room temperature. It is important to keep the relative ambient air humidity below 40 %. Small samples cut out of lyophilized cryostat sections can be weighed at room temperature on a Lowry Quartz Fiber Balance [4,5]. If frozen specimens are to be homogenized, weighing can be omitted. Instead, part of the homogenate is pipetted off for subsequent quantification. Suitable methods are available for the required microanalysis and examples are given below.

Small amounts of protein are determinable in fluorometric assays by the fluorescamine methods [6]. A procedure for measuring quantities between 0.5 and 10 μg has recently been described in connection with analyses of adenylates [7], as shown in Figure 9-17. Proteins can also be determined with a spot technique. In one method quantities in the ng range are measurable fluorometrically after staining with a benzoxanthene derivative (Hoechst 2495). After staining a spot applied to a cellulose acetate strip, the spot and strip may be dissolved together in dimethylsulfoxide for convenient analysis [8,9].

For DNA quantification, modifications of a fluorometric method based on a reaction with deoxyribose have turned out to be particularly useful [10,11]. An important advantage of the method is that the presence of protein is tolerable. Protein may thus be used as a protective agent against surface inactivation, if required. Extraction of cells with saponin spoils subsequent DNA assay, probably due to interference with its glucoside component. In our laboratory the procedure described in Figure 6-4 is used for quantification:

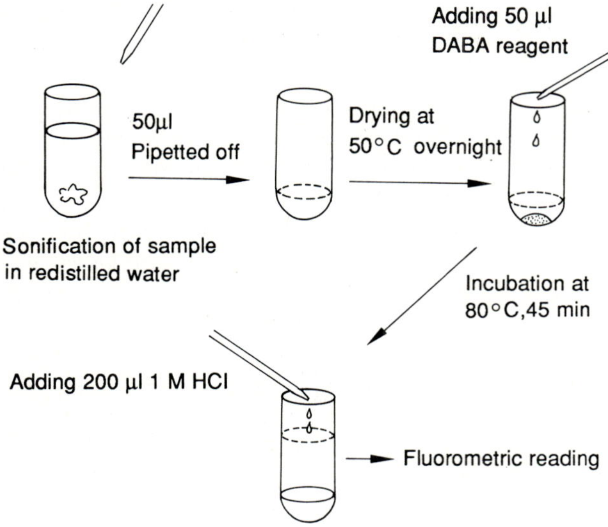

Figure 6-4. Quantification of cell and tissue samples, based on fluorometric determination of DNA.

DABA reagent: A fresh solution of 3,5-diaminobenzoic acid dihydrochloride in 1 500 μL redistilled water is decolorized by shaking at least three times with 10–20 mg active carbon.

Standards: Stock solution: 10 mg DNA from calf thymus (Sigma D1501) is dissolved in 50 μL 1 M NH_4OH and may then be stored refrigerated for months. Starting with 100 μL stock solution added to 900 μL NH_4OH, standards are freshly prepared by serial dilution with the ammonia solution (500 μL + 500 μL) down to 0.3125 μg DNA per 500 μL. Droplets of 50 μL standard solution are deposited in the same way as samples and blanks at the bottom of test tubes and dried overnight.

Blanks: 50 μL 1 M NH_4OH

A Farrand fluorometer may be employed, using as primary filter No. 205 and secondaries Nos. 2A–12 and 65A.

The above-mentioned techniques are all well-established. However, there appears to be a need for a convenient means for quantifying minute cell samples, where the above methods may have their difficulties. In the section on biomass determinations (Chapter 8), the applicability of nucleotide analyses for quantification purposes is discussed. Attention is drawn to the possibility of using the sum of nucleotides (ATP + ADP + AMP) for quantifying small amounts of mammalian cellular material. Some aspects of the results of various quantification measures, as the estimate of biomass in different cell populations, are discussed in Section 8.2 and Figure 8-5.

6.4 Planning the Initial Preparative Steps

6.4.1 Sampling from Cell Suspensions, Cultures of Cells and Small Organs

In the preparation of cultured cells and organs for bioluminescence analysis, a safe and reliable sequence of preparative steps can be distinguished and is marked by coarse arrows in Figure 6-5. Fine arrows mark alternative routes or complementary procedures. Dotted arrows indicate quantification steps.

Harvesting is the first preparative step. It is normally followed by pelleting and freezing. For quantification purposes, cell counts may be carried out parallel to the harvesting. To ensure that representative samples are formed and also to collect information about the functioning of the cells, checks on the condition of the medium may be required (route 1). The results may call for washing procedures before the cells can be pelleted. In the pellet, the situation may resemble ischemia of an organ and can be associated with biochemical changes observed in such states. Thus, in order to minimize any disturbances of this kind, it is important that all operations, if possible, are carried out at low temperature, and that the time between harvesting and freezing is short. For instance, centrifugation to form a pellet should preferably take place in a refrigerated room. With small organs, pelleting may be omitted and the organ directly frozen (route 2). This is described above (Section 6.2.1) and involves rapid transfer of the organ into chilled isopentane or freon.

A few enzymes are susceptible to freezing damage and have to be analyzed directly, as indicated by route 3. This route involves homogenization in a hypotonic buffer with or without subsequent addition of detergents. It is difficult to carry out other procedures and analyses in addition to the enzyme assay. With this type of preparation, nucleotides and substrates are generally converted to an unacceptable extent. Alternatively, as a modification of route 3, lysis of cells may be accomplished in the culture medium by adding a suitable detergent or by cautious sonification, making several preparative steps unecessary (cf. Figures 9-13, 9-14, and 9-15). Prior addition of luciferase and reactants to the culture medium makes their biochemical content principally determinable, but applications require careful calibration and attention to degradation that may accompany this kind of release.

To provide for convenient storage and handling, freezing is normally accompanied by lyophilization. This yields specimens which can be dealt with and weighed at room

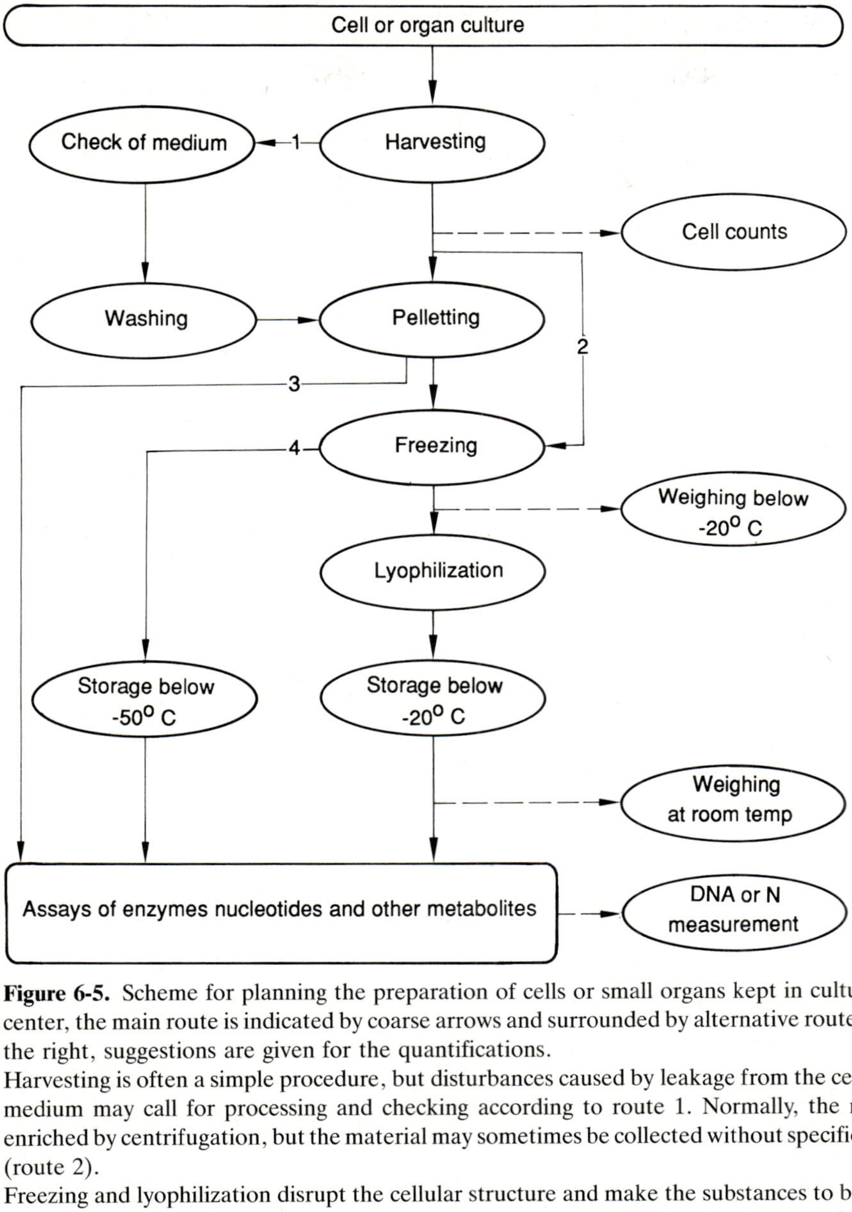

Figure 6-5. Scheme for planning the preparation of cells or small organs kept in culture. In the center, the main route is indicated by coarse arrows and surrounded by alternative routes (1–4). To the right, suggestions are given for the quantifications.
Harvesting is often a simple procedure, but disturbances caused by leakage from the cells into the medium may call for processing and checking according to route 1. Normally, the material is enriched by centrifugation, but the material may sometimes be collected without specific measures (route 2).
Freezing and lyophilization disrupt the cellular structure and make the substances to be analyzed readily accessible. Routes 3 and 4 point to simplifications. Following homogenization, the determination of enzyme activity can usually be accomplished. For analyses of nucleotides and other metabolites, route 3 is not advisable without checking the extraction yield and excluding major losses through degradation or conversion of the compound to be determined.
Extraction procedures and final processing for analysis are shown in Figures 6-7, 6-9 and 6-10.

temperature (cf. Section 6.2.2). When preparing lyophilized small organs for weighing, handled according to route 2 (Figure 6-5), it is normally necessary to peel off remnants of the culture medium which appear as a fluffy coat on the samples after lyophilization.

This process is preferably carried out under a stereomicroscope. Lyophilized specimens require only −20 °C in long-term storage, as compared to frozen samples which need −50 °C or below (route 4). This route, however, implies awkward weighing conditions, but quantifications can be based on cell counts. Furthermore, quantifications based on DNA or nitrogen determinations are generally applicable, provided that additives do not interfere.

6.4.2 Sampling from Biopsies and Pieces of Organs

A scheme for preparing organs for analysis is shown in Figure 6-6. The main route is partly similar to that of Figure 6-5. Cell counts are out of the question here. Otherwise,

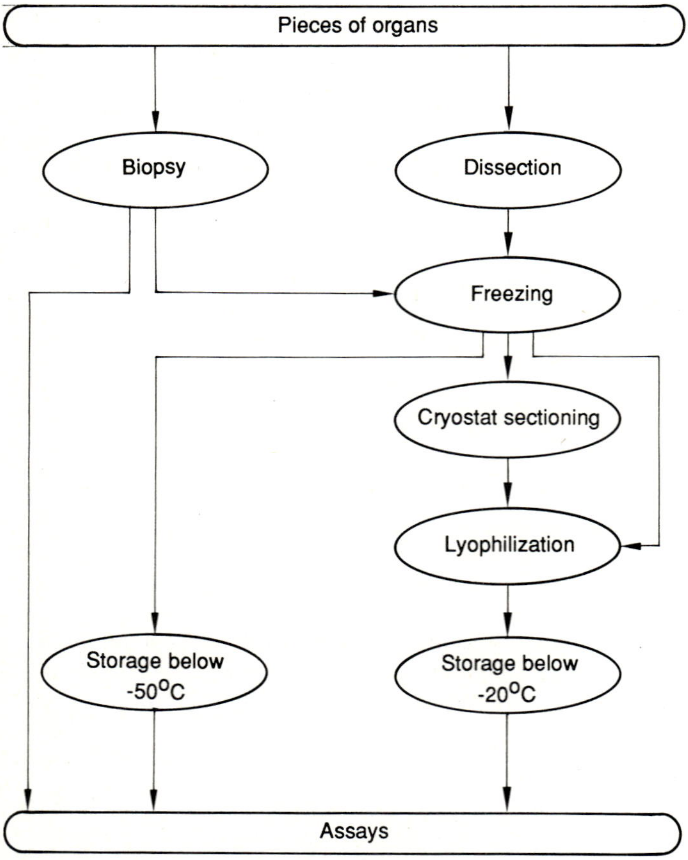

Figure 6-6. Planning the preparation of organ specimens. Samples may be obtained from human biopsies or from experimental animals. Testing with animals can decide whether instantaneous freezing is needed for a satisfactory metabolic arrest. Except for cell counts which are irrelevant in this context, the requirements for quantification are indicated in Figure 6-5.
It is sometimes possible to do without the cryostat technique as indicated on the right of the scheme. The simplifications shown on the left are well practicable in enzyme assay, but applications for metabolites require testing for essentially maintained concentrations, preferably by comparison with samples processed according to the main preparative route.

the opportunities and requirements for quantification are in agreement with those indicated in Figure 6-5.

With organs, lyophilization provides the additional advantage of permitting the dissection of defined cell groups at room temperature, using cryostat sections, cf. Figure 8-7. If sectioning and lyophilization are omitted, extraction procedures become more critical. Powdering the specimen by a microdisintegrator in liquid nitrogen facilitates subsequent processing, but it should be kept in mind that the water bound to the tissue may be an obstacle at the extraction and may call for particular attention.

How a metabolic state can be preserved by very rapid freezing of the extirpated organ is described above (Figures 6-1 and 6-2). Such procedures may be suitable for animal research, where dissection can be carried out freely under anaesthesia. Application of biopsy techniques, which can be extended to human tissues, generally requires longer times for handling before the sample is safely preserved in the frozen state. It appears, however, that handling times on the order of a few minutes are often also permissible in the analysis of metabolites and nucleotides, which in other cells are known to undergo changes, when the blood supply has been interrupted. One application has been to the superficial layers of the skin, where biopsy techniques were successfully combined with luminescence analyses and the effect of delayed freezing was evaluated [12].

As discussed above, enzyme assays are not critically dependent on very rapid freezing, which may even be contraindicated. Sometimes, very rapid freezing can be unnecessary, leaving cells that have been isolated from biopsies to recover in a culture medium before further processing. This has been carried out in a clinical study on differences in lipolysis, using bioluminescence analysis of glycerol release [13,14].

6.5 Extraction Methods

6.5.1 General Procedures

As described in the preceding sections, the initial preparative steps are designed to allow an interval, during which series of samples can be collected and then extracted and processed ready for analysis, either directly or after storage. Extraction is generally carried out using various combinations of mechanical disintegration and chemical action. Either option may be more or less predominant in a particular procedure. Among mechanical measures, disruption of cells in a glass blender is frequently employed but can be replaced by cautious sonification. Hypotonicity as well as freezing break cellular membranes, favoring extraction. Freeze-clamped tissue samples can be powdered in liquid nitrogen with a microdisintegrator, increasing the surface contact for subsequent extraction. In samples dissected from lyophilized cryostat sections the removal of water in connection with capillary forces facilitate the extraction. Since cryostat sectioning already reduces the samples to small dimensions, further mechanical processes are unnecessary. A word of caution: Vibrations of microtubes or microhomogenization may cause part of a sample to stick to the wall of the tube or homogenizer and thus introduce errors.

The choice of chemicals for the extraction is mainly determined by the resistance of the cellular membrane. Mammalian and microbial cells show a great difference in this respect. When they occur together, their content can be estimated separately by starting with a mild extraction for disrupting mammalian cells, followed by addition of a more potent extractant for the lysis of microbial cells [2,3]. This technique has been applied to the diagnosis of bacteruria, Section 8.2.2.

In the extraction of mammalian cells, various detergents may serve as membrane destroyers. Their properties have been discussed in connection with the technique employed for the extraction of tubular organs, see Section 9.5.3. Here it may suffice to draw attention to the detergent compound G3707 (heptaoxyethylene lauryl ether). It has been found to be useful in preparing samples for analyses of both metabolites and enzymes and also to permit the determination of DNA.

However, unicellular cultures and suspensions also allow a choice of extraction method for the substance to be analyzed. Thus, the extraction of one nucleotide form may call for a different method than the extraction of another. Furthermore, in some analyses, interconversion between the various nucleotide forms may be permitted, whereas in other cases it is important that the metabolic contents are kept unaltered. Thus, when the total amount of adenine nucleotides is used for biomass determinations, the demand on the extraction is much less rigorous than when a separate determination of the contents of ATP, ADP and AMP is desired. In the design of the preparative schemes presented in this chapter, preference was given to established methods, but it should be kept in mind that particular analytical situations may call for modification of these methods. In cryobiology, elaborate methods have been developed for preserving the viability of frozen cells. These sometimes complicated procedures naturally permit the application of bioluminescence analysis, but do not provide any advantage, if the frozen cells have to be subject to disruption and extraction when processed for analysis.

6.5.2 Adenine Nucleotides

For the extraction of mammalian cells, disintegration by hypotonicity or mechanical means may often be sufficient. Furthermore, various detergents may serve as membrane destroyers. Some of their properties have been discussed in connection with the technique employed for studying discharge from cells and extraction of tubular organs, cf. Sections 9.5.3 and 9.8.

Extraction of microbial cells is often carried out by treating the cells with trichloroacetic or perchloric acid, followed by centrifugation. This results in a protein free solution. The acid will inactivate any enzyme, including added luciferase, and must be diluted or preferably removed before bioluminescence analysis. Trichloroacetic acid can be extracted with water-saturated diethylether, and the main portion of perchloric acid precipitates as its potassium salt. High extraction yields of adenine nucleotides are reported. In practical applications, such as in the diagnosis of bacteruria, it is preferable to use extractants which need not be removed before analysis. Whether the yield remains sufficient can be checked by comparison with trichloroacetic acid extraction [2,3].

For studies of metabolic conditions in living mammalian cells, certain extraction procedures have been used, which are based on experience with a great number of fluorometric and bioluminescence assays. The procedures in Figure 6-7 concern samples

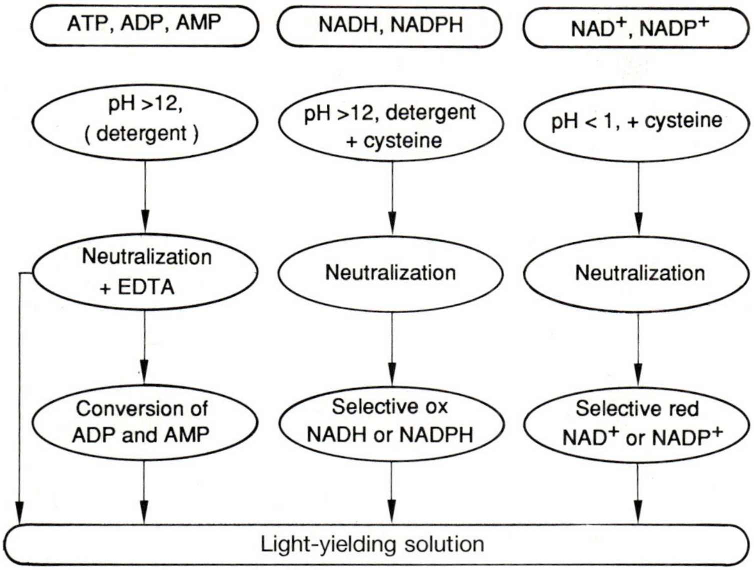

Figure 6-7. Completion of the sample processing for analyses of nucleotides.
Extraction at either high or low pH values prevents disturbing conversions in enzyme reactions. After preceding lyophilization (cf. Figures 6-5 and 6-6), the yield of nucleotides is generally satisfactory but may be too low when this measure is omitted. In this case a detergent may be added to enhance the extraction. This applies likewise to cells with resistant membranes.
Chelating Mg^{2+} with EDTA prevents undesired interconversions in the adenylate pool. Cysteine protects against undue oxidation, but not at low pH. Neutralization involves suitable dilution in order to minimize interference by disturbing sample components (see Figure 6-3).
The alternatives of either determining ATP alone or the adenylate sum are indicated on the left.
The possibilities of destroying either the oxidized or the reduced forms of pyridine nucleotides are discussed in Figure 6-8. The four forms can be determined after selective oxidation (ox) or reduction (red), as shown at the bottom of the diagram. Using enzymatic cycles, continuous reduction in the light-producing solution provides an other alternative to specific determinations (cf. Figure 7-1). Except for the final nucleotide conversions, the processing is carried out on ice.

with arrested metabolism, following the schemes outlined in Figures 6-5 and 6-6. Conversion between the nucleotide forms during extraction can be prevented by eliminating the catalytic action of enzymes in the sample. In principle, interfering phosphotransferases are at least partly inactivated and their action interrupted at extremely high or low pH values. Aqueous solutions of 0.05 or 0.02 M NaOH have been used. The extraction (cf. Figure 6-7) may start at 0 to −4 °C with a glass homogenizer or sonification, but small groups of cells isolated from lyophilized cryostat sections need only to be inserted into the extraction fluid. When a blender or vibrator is used, detergent must be added afterwards, because of the resulting foaming. The extraction is completed by heating at 60 °C for 15 min. After neutralization, interconversions may nevertheless begin in the adenine nucleotide pool but are precluded by dilution and addition of EDTA acting as a chelator of Mg^{2+}. In case the presence of phosphatases is disturbing, addition of vanadate or ouabain may be expedient, since they have been known to inhibit ATPases without affecting the firefly luciferase system [15].

6.5.3 Pyridine Nucleotides

In the biological transformation of energy, NAD^+ is the main oxidant. Accordingly, the conversion of NAD^+ to NADH includes many steps in the degradation of metabolites, providing for fast interconversion between oxidized and reduced forms. In this way, the metabolic state of the cell is related to the quantities of these forms, rendering it essential to efficiently block interconversion and to be able to analyze them separately. This is a considerably more difficult venture than determining the sum of the nucleotides. Figure 6-7 indicates the preferred approach. The analytical procedures apply likewise to $NADP^+$ and NADPH. Interconversion between these species, which are specifically involved in many synthetic steps of the cell metabolism, is generally slower.

Widely different pH values are the basis for separate extraction of either NAD(P)H or $NAD(P)^+$. This is in agreement with selective destruction of the reduced forms with acid and the oxidized nucleotides with alkali, Figure 6-8. By employing a very short

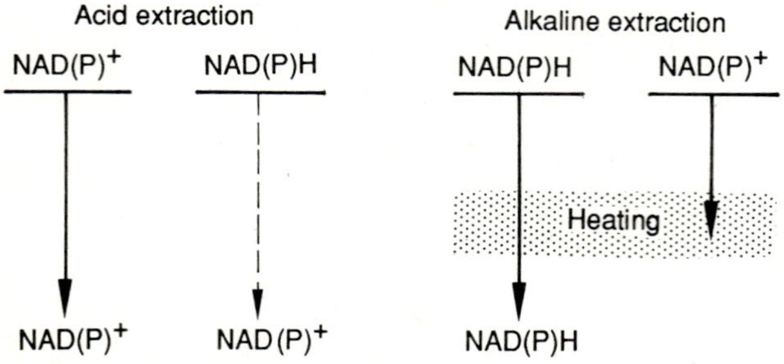

Figure 6-8. Selective extraction and corresponding destruction of the oxidized and reduced forms of pyridine nucleotides at extreme pH values.
Acidification (pH<1) allows the extraction of oxidized nucleotides, whereas alkaline solutions (pH>12) extract the reduced forms [4]. Particularly at low pH values, the presence of hemoglobin may lead to the oxidation of reduced nucleotides and hence increase the extraction yield (dashed arrow). Alkaline extraction of lyophilized specimens combined with heating is generally safe and reliable.

extraction time in alkali, e.g. not more than one minute, it is possible to leach out the reduced forms without causing significant destruction of the oxidized form and thus estimate the concentration of both forms. In general, cysteine is added to the extracting solution as an antioxidant and EDTA as a chelating agent for metals. Any still active, metal-dependent enzymes are thus forced into inactivity.

The separate extraction of either the reduced or the oxidized pyridine nucleotides is reliable when it concerns cells in a culture or such avascular samples as skin biopsies. The presence of hemoglobin may provoke a shift between the reduced and oxidized nucleotide forms, as the protection from cysteine may become inadequate, so that part of the NAD(P)H is oxidized. This effect is most pronounced in an acidic environment but may be present also in alkali, if the hemoglobin concentration is very high. In addition to cysteine, which may be insufficient, ascorbate is also used to counteract oxidation during acid extraction [4,16]. Ascorbic acid is a comparatively strong acid and may interfere with the pH, but also disturb the luciferase reaction in other ways. The extraction may be started with a minor portion of the acid solution supplemented with 2 mM ascorbic acid. However, before entering into the problems of undesired oxidation mediated by hemoglobin in an assay, its presence should be demonstrated and the quantitiy estimated. Suitable and sensitive bioluminescence methods meet these demands (cf. Section 9.5.2).

As mentioned above, short alkaline extraction on ice, followed by neutralization, is useful as a preparatory step for analyses of all different forms of pyridine nucleotides; but the inconvenience of checking the preparation times of the single samples in a series may call for alternative procedures. For analyses of NAD^+ and NADH in freeze-clamped tissue samples, a simultaneous extraction was performed using buffered ethanol [17]. Without preceding lyophilization as needed in alkaline extraction, a high yield of both nucleotide forms was obtained. Furthermore, a mixture containing phenol was employed for common and convenient extraction of the pyridine nucleotide forms [18]. This method is likewise applicable when it comes to subsequent bioluminescence analysis, provided that the problem of disturbing oxidation is kept in mind.

6.5.4 Intermediary Metabolites

There is a significant difference between assays of intermediary metabolites and nucleotides. In the analysis of metabolites, a specific enzyme is added, together with a nucleotide which serves as a cofactor during the conversion to its light-inducing form. The nucleotide is generally supplied in excess, which means that interfering reactions may increase the amount of converted nucleotide and thereby the blank readings. Because of this, sample processing is much more critical than in the determination of nucleotides by converting them to light-inducing forms by analogous addition of an enzyme and an excess of metabolite.

Most intermediary metabolites can be extracted with aqueous solutions, particularly after preceding lyophilization. The processing for analysis may therefore start with homogenization in a cold hypotonic buffer setting the pH for optimal yield of the target compound, see Figure 6-9. The aforementioned disturbances, resulting in elevated blank readings, have been ascribed to micro particles and molecular aggregates suspended in the solution. At ordinary centrifugal forces, they remain in the supernatant [19], but disturbing emissions can easily be reduced by high-speed centrifugation with an airfuge in a refrigerated room. Speed and time of centrifugation are adjusted to decrease blank readings.

High-speed centrifugation may be replaced by protein precipitation, but the precipitant must not disturb the light production by inactivating luciferase. In our laboratory, deproteinization according to Somogy [20] has been found suitable for the processing of samples. In principle, zinc sulfate and barium hydroxide are added in amounts adjusted such that barium sulfate and zinc hydroxide precipitate almost completely. A small excess of either zinc or barium can be chelated with EDTA.

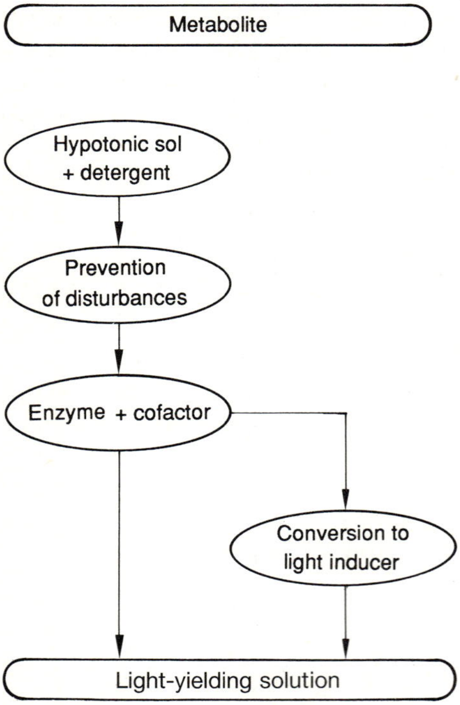

Figure 6-9. Processing of samples for analysis of metabolites.
Samples prepared according to Figures 6-5 and 6-6 are homogenized in a cold hypotonic buffer. Detergents enhance the extraction but cause foaming and must not be added before homogenization.
Prevention of disturbances largely implies the use of highspeed centrifugation or precipitation to remove microparticles and aggregates of molecules. Such measures are particularly needed in analyses based on the conversion of excess amounts of pyridine nucleotide.
The processing is completed by adding the specific cofactor and enzyme for the metabolite to be assayed. With pyridine nucleotides, NAD(P)H is the light inducer and directly measurable. Measurements of formed $NAD(P)^+$ require reduction after selective destruction of added NAD(P)H.

6.5.5 Enzymes

Most enzymes may be preserved for subsequent analysis, as described in the planning for the initial preparative steps, Section 6.4. The processing depends to a large extent on the size of the sample. Small samples, usually in the ng range, are obtained when defined groups of cells are cut out of cryostat sections (cf. Figure 8-7). These specimens are inserted in a buffer substrate designed for the production of a light-inducing compound. Mixing should be avoided; otherwise, the sample may adhere to the tube wall. Other kinds of specimens, which are larger, are preferably homogenized in a cold hypotonic buffer, either after lyophilization or simply storage at a low temperature. A suitable amount of tissue may vary between 5 and 25 mg dry weight per ml, providing a sufficient amount of protein for protection against enzyme inactivation. However, dilution may be desired during preparation in order to diminish disturbances due to side reactions. Further addition of protein may therefore be advisable (cf. Section 6.2.4). It is difficult to estimate to what extent the disruption of cellular structures by freezing improves the analyses, but a few enzymes have a tendency to be damaged by this process. In this case, homogenization of fresh samples is necessary.

A simple scheme which is generally followed in enzyme assays is shown in Figure 6-10. The final addition of substrate and cofactors produces a light-inducing compound for

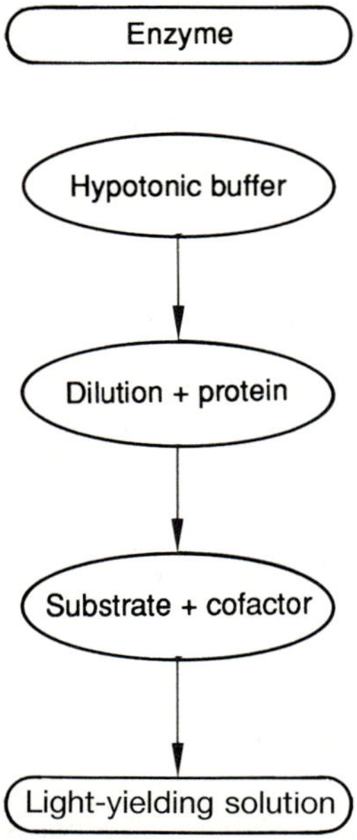

Figure 6-10. Processing of samples for analysis of enzymes.
The scheme is applicable to enzymes prepared as suggested in Figures 6-5 and 6-6. It starts with homogenization in a cold hypotonic buffer. Frequently, it is also possible to start with a homogenate of a fresh piece of organ.
To remove disturbances, a high dilution may be required (cf. Figure 6-3) – permitted by the high sensitivity of bioluminescence in connection with the chemical amplification obtained by product accumulation when determining enzyme activities. Enzyme inactivation caused by dilution has to be counteracted by addition of protein.
In the last step, a specific substrate is added, together with a cofactor, which is converted to its light-inducing form. It is alternatively possible to supplement the luciferase solution with these additives and to read the maximum emission, which in reality represents a close approach to the equilibrium displayed in Figure 4-6.

subsequent measurement. As already mentioned above, this procedure is employed when pieces dissected from cryostat sections are analyzed.

The same procedure applies to samples pipetted from diluted homogenates, but these may be added directly to a luciferase solution, which contains the additives required for the production of light. In this case, the assay may be arranged such that an emission equilibrium is approached at which the intensity stays constant even with partial luciferase inactivation, hence reducing analytical errors (cf. Figure 4-7).

References

[1] Wollenberger A, Ristau O, Schoffa G. Pfluergers Arch Gesamte Physiol. 1960; 270:399–412.
[2] Lundin A, Thore A. Appl Microbiol. 1975; 30:713–721.
[3] Stanley PE. In: DeLuca MA, McElroy WD, ed. Methods in Enzymology New York; Academic Press; 1986; 14–21. (Bioluminescence and Chemiluminescence, Part B; vol 133).
[4] Lowry OH, Passonneau JV. A Flexible System of Enzymatic Analysis. New York: Academic Press; 1972.
[5] Lowry OH. J Histochem Cytochem. 1953; 1:420–428.
[6] Udenfriend S, Stein S, Böhlen P, Dairman W, Leimgruber W, Weigele M. Science 1972; 178:871–872.
[7] Brolin SE, Naeser P, Hjertén S. Prep Biochem. 1987; 17:327–336.
[8] Neuhoff V, Philipp K, Zimmer H-G, Mesecke S. Hoppe-Seyler's Z Physiol Chem. 1979; 360:1657–1670.
[9] Neuhoff V. In: Michal F, ed. Application of Biochemical Micromethods for the Investigation of Tropical Disease Pathogens. Geneva: UNDP/WORLDBANK/WHO; 1982; 39–110.
[10] Kissane JM, Robins E. J Biol Chem. 1958; 233:184–188.
[11] Hinegardner RT. Anal Biochem. 1971; 39:197–201.
[12] Hammar H. Acta Dermatovener. 1973; 53:251–258.
[13] Kather H, Schröder F, Simon B. In: Serio M, Pazzagli M, eds. Luminescent Assays Perspectives in Endocrinology and Clinical Chemistry. New York: Raven Press; 1982: 53–56.
[14] Kather H, Wieland E, Scheurer A, Vogel G, Widenberg U, Joost U. J Clin Invest. 1987; 80:566–572.
[15] Baltus E, Hanocq-Quertier J, Schram E. In: Schölmerich J, Andreesen R, Kapp A, Ernst M, Wood WG, eds. Bioluminescence and Chemiluminescence New Perspectives. Chichester: John Wiley & Sons; 1987: 567–570.
[16] Burch HB, Bradley ME, Lowry OH. J Biol Chem 1967; 242:4546–4554.
[17] Karp MT, Raunio PR, Lövgren T. Anal Biochem. 1983; 128:175–180.
[18] Neuhoff V, Desselberger E. Naunyn-Schmiedebergers Arch Exp Path Pharm. 1965; 252:43–49.
[19] Brolin SE, Naeser P, Berggren P-O. In: Kricka LJ, Stanley PE, Thorpe GHG, Whitehead TP, eds. Analytical Applications of Bioluminescence and Chemiluminescence. New York: Academic Press; 1984: 515–519.
[20] Somogy M. J Biol Chem. 1945; 160:69–73.

7 Generation and Measurement of Light Signals

7.1 Paths Towards Analytical Response

The practical performance of an analytical luminescence recording is basically a question of separating the relevant light signal from disturbing signals. It is a matter of enhancing the reaction steps involved in the desired analytical pathway and at the same time suppressing disturbing reactions. It is also a matter of selecting the instrumentation and, in particular, the light-measuring devices which can be adjusted to record the specific signal determined by the analytical arrangement.

In the preceding chapters, it is described how systems can be set up for analyses of various substrates and enzymes and coupled to a final step, giving rise to the emission of light. The reagents can in principle be purified in order to eliminate interfering reactions caused by reagent contaminants; and steps can be taken to increase the stability of the target enzymes. Encapsulation and immobilization of enzymes are generally useful to produce preparations which are well suited for storage, and such techniques have been successfully applied to the firefly [1–3] and the bacterial luciferase systems [3–5]. These, however, lack some of the advantages of ordinary homogeneous enzyme solutions, but are less likely to be inactivated during the assay and are thus capable of maintaining a low conversion rate. Reutilization is possible and evidently of great value in flow systems.

Highly purified enzymes are nowadays available, so that side reactions due to impurities are virtually eliminated. With the samples to be analyzed, the situation is more difficult and troublesome. Precision and reliability in isolating the target compound may be difficult to attain, in particular in the case of small samples. Instead, various steps have to be taken to minimize disturbances by unwanted components. In other words, while it may be desirable to attempt to specifically extract the compound to be measured, it may be more important to design procedures that minimize interfering reactions. As emphasized in the preceding chapter, the means used in the preparation and processing of the sample are important, and so is the way the luminescence reactions are set up in the final analyses. An example of the close interdependence is as follows. During preparation and sample processing, a pH shift may have been used to prevent an unwanted reaction. However, it may not be possible to maintain this shift of pH during light production. Therefore, other steps must be taken to minimize the influence of such reactions.

Experience with diminishing the influence of side reactions has to a large extent been collected in the course of experiments with crude extracts from luminescent organisms. Since purification without losing a major portion of the enzyme activity was difficult, it was preferable to design methods which kinetically eliminated unwanted reactions. Errors caused by sample components can largely be eliminated and compensated by the same methods.

7.2 Desired Emission by Kinetic Means

7.2.1 Flash Techniques

Flash techniques proved early to be a powerful means of decreasing the influence of erroneous emission, Table 7-1. It is a matter of enhancing the rate of the desired

Table 7-1. Measures taken to decrease disturbing emission in different analytical arrangements

Emission type	Measures
Light flash or early maximum appearance	Dilution[1] Getting above and ahead of disturbing interference when recording maximum display For coupled ractions: Design for kinetic compensation of luciferase inactivation
Long-lasting emission due to slow consumption of the light-producing substance	Compensation for tolerable deviations by using internal standards Using a setup with raised emission intensity permitting sufficient dilution
Stable emission maintained by continuous reconstitution of light-producing substance	Allowing disturbing emission to fade out, due to consumption of contaminating light-producing substances Design for kinetic compensation of luciferase inactivation

[1] Removal of disturbances by dilution begins with the processing of the samples and is completed with the selection of sample volumes and the light-producing solution

analytical reaction, so that its emission becomes clearly distinguishable as a light flash above the background level produced by other reactions and by the light detection system (cf. Figures 5-6 and 5-9). The technique of high sample dilution is often applicable. Unwanted reactions are slowed down by diluting the samples, while the desired reaction is enhanced by adding high concentrations of luciferase and auxilliary components. Small samples are also used to the same effect (Figure 6-3). This approach is based on the availability of a sensitive detection system with inherent low noise, as the technique also implies a reduction of both the concentration of the substance to be determined and of the signal level. Reproducible rapid mixing is also necessary with the flash techniques. Further discussion of the demands on the experimental equipment is given in later paragraphs of this chapter.

7.2.2 Delayed Appearance of the Emission Maximum

The advantages of using a flash technique are also largely evident in cases where the emission is merely a delayed appearance of a maximum rather than a true flash of light. A light flash is characterized by a maximum emission that occurs within a few seconds after initiating the reaction and that is followed by rapid fading. By allowing for a somewhat later appearance of the maximum, it is possible to make handling easier and still be sufficiently ahead of interfering reactions. Manual mixing and positioning of the reaction vessel are possible. Attention should be paid to the fact that the coupled recations have the great advantage of a high-concentration luciferase step in connection with a slower preceding enzyme reaction. The system is described in more detail in Chapters 4 and 5. Here, the influence of impaired luciferase activity during the course of a series of analyses is essentially eliminated. This facilitates reproduction and simplifies calibration.

When the rising time of the luminescence emission is to be manipulated, it is necessary to be able to select the rate of the reaction preceding the light-producing step. In other words, luminescence analysis has to be a set of two consecutive reactions for which appropriate rates can be selected. With the luciferase system used for the analysis of ATP, there exist no appropriate means of varying the rising time of the emission, but controlled addition of inhibitors may offer a possibility. A delayed appearance of the emission maximum may be possible, if the system is used together with a preceding reaction, e.g., for the analysis of ADP or such compounds which phosphorylate ADP. The situation is different for the bacterial luciferase system used for the analysis of NAD(P)H, which consists of two coupled reactions. In this case, the rates of the two steps, the oxidoreductase and the luciferase reaction, can be individually controlled, see Figures 5-8 and 5-9. An additional preceding step in the form of a dehydrogenase reaction makes the method of measuring the delayed appearance of an emission maximum applicable to a large number of metabolite and enzyme analyses.

7.2.3 Long-Lasting Emission

Arrangements for measuring intensity changes of reagents that lead to durable emissions have been found to be attractive for the analysis of ATP and coupling to ATP formation. Such systems have been discussed in Chapter 5, where it is also mentioned how disturbing reactions may be detected. Internal standards are recommended. It is, however, often satisfactory to certify agreement between a few internal and ordinary standards. cf. Section 5.2.3. As with all luminescence analysis systems, the reagent should have a low background as compared to the intensity levels reached when samples are added. The light emission response should, furthermore, remain unimpaired during the time required to perform a series of analyses. For analyses of adenylates, the general experience with this kind of system is positive. There are only exceptional cases in which disturbing components of a sample rule out application of this technique (Table 7-1).

7.2.4 Stable Emission

The systems commonly used for durable intensity readings are in reality slowly fading. More truly constant signal systems may be obtained by coupled reactions involving internal reconstitution of a light-inducing reagent. Such systems may also compensate well for losses in luciferase activity, i.e., essentially the same intensity level may be obtained, even if the luciferase is partly inactivated during the course of an analysis. How this is achieved through setting appropriate rates for the coupled reactions is described in Chapter 4. The overall signal level and thus the sensitivity may be increased by raising the activity of all enzymes required for the coupled reaction and letting the individual steps proceed at about the same rate.

Most applications concern the bacterial luciferase system. Its great advantage is the fact that reconstitution opens up the possibility of distinguishing between various nucleotide forms. Hence, selecting and coupling to a dehydrogenase reaction which is specific for one nucleotide has the effect that only this form is regenerated. After consuming any other nucleotide during the initial period of the analysis, the light signal will ultimately approach a constant level defined only by the regenerable nucleotide to be determined, see Figure 7-1. It should be recalled that the cycling measures the total nucleotide amount, i.e., the sum of the reduced and the oxidized form. To determine just one nucleotide form, reduced or oxidized, it is necessary to selectively destroy the other form prior to the analysis.

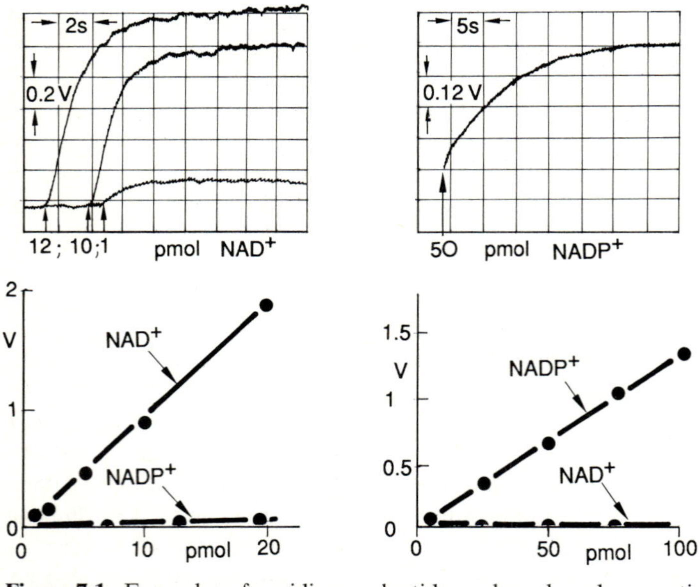

Figure 7-1. Examples of pyridine nucleotide analyses based on continuous reduction.
Left oscillogram: Different amounts of NAD^+ (arrowed) are injected into a solution of partly purified bacterial luciferase, malate dehydrogenase and malate. The detailed composition is given in [6]. The overall reaction rate is kept high enough to rule out interference by contaminants. Time is allowed to let erroneous initial disturbing emission fade out before signals are read. Such disturbances may have affected the lowest curve. The calibration below gives evidence for linearity and specificity. The signals from the $NADP^+$ samples are negligible and are possible caused by traces of NAD^+, contaminating the samples.
Right oscillogram: Approach to equilibrium after injection of a $NADP^+$ sample into a luciferase solution containing glucose-6-phosphate dehydrogenase and its substrate [7]. Oxidoreductase and luciferase are extracted and purified from crude bacterial luciferase extracts [8]. Specificity and a wide range of linearity are shown below by the calibration.
It can be seen that the two analyses are carried out with about a tenfold difference in sensitivity. They therefore illustrate that the reaction rates can be set to match the measuring range of the luminometer, the amount of nucleotide to be assayed, and the kinetic possibilities of diminishing disturbing interference.

A specific example is the determination of NADP(H) by coupling to the bacterial luciferase system, using the glucose-6-phosphate dehydrogenase reaction. This means that any NADH present or formed will not be reconstituted by the enzyme. In practice, it may be an advantage to start the reaction using a solution without glucose-6-phosphate. An estimate of the presence of reduced nucleotide (NADPH and NADH) is obtained from the resulting light signal which fades. With glucose-6-phosphate added a steady signal level is approached, and this indicates the combined amount of $NADP^+$ and NADPH. To determine one of the two nucleotides separately, the other must be selectively destroyed in advance, either by acidification or by treatment with alkali. Many metabolites can be measured by letting them react specifically with NADH or NADPH and estimating the amount of oxidized nucleotide. The oxidation is interrupted by acidification at a fixed time, which meanwhile also destroys the reduced nucleotide, leaving the oxidized form available for measurement.

7.3 The Light-Measuring Procedure

7.3.1 Light Recording in Photometric Methods

When discussing the light-measuring procedures and the sample processing involved in chemiluminescence analysis, comparisons with absorption spectrophotometry and fluorometry are often instructive, because these represent the two standard techniques for quantification in a chemical or biological laboratory.

Virtually all compounds absorb light and can thus be determined by spectrophotometric means. However, absorption and light scattering of the other species in the test solution pose strong limitations with respect to both specificity and sensitivity. Furthermore, the technique implies measurement of the light intensity before and after absorption, using a monochromatic beam. Altogether, this means that the limit of detectability lies in the ppm region and that turbidity must be low. The recording of light transmission allows accurate measurements only within a small concentration range, optimally when the transmission is between 10 and 90 %. To cover a large concentration range, it is therefore necessary to dilute some samples.

Relatively few compounds fluoresce, and interference by emitting species other than the substance to be determined is less likely in fluorescence analysis than is a disturbing absorption due to other species during absorption spectrophotometry. There may, however, be components of the sample capable of quenching the excited state responsible for the fluorescence, weakening the emission. Interference from a turbid solution is, on the other hand, less of a problem than in absorption spectrophotometry. Only one signal, the fluorescence intensity, has to be measured, but the result is directly related to the stability of the exciting beam. A remarkable concentration range may be covered with a linear response over 4–6 orders of magnitude. The ppb and even the sub ppb region may be reached.

Chemiluminescence or bioluminescence is a much more rare phenomenon than are light absorption and fluorescence. The chemical reaction giving rise to light may, however, be interfered with in various ways. Thus, like other enzymes, the different luciferases are susceptible to inhibition and degradation. However, many situations allow kinetic compensations for this kind of disturbance. A more specific problem of luminescence analysis is the interference with the excited species, leading to possible quenching of the light emission by various mechanisms. Such processes may give rise to uncertainties in the analytical response.

Since there is no incoming light beam in chemiluminescence analysis, there is a greater versatility with respect to selecting sample shape and size. With an incoming beam, the analysis is based on an illuminated volume in the sample cuvette, whose concentration attenuates the beam or gives rise to emission. In luminescence analysis, the ultimate sensitivity does not only depend on the number of molecules involved, but also on the reaction rate, determined by the concentration. If the amount of reacting species remains constant and the reaction volume decreases, the intensity may rise because of the concentration increase. We can conclude that the required minimum size for sample handling and the efficiency of collecting and recording the light output are very important factors in defining the ultimate sensitivity.

Luminescence analysis brings about some electronic background noise, but in principle this can be kept at a very low level; and with a decent quantum yield, one can theoretically think of possibilities where the number of analyte molecules in a sample are reacted and "counted". As with fluorescence analysis there is usually a very large concentration range over which the method is linear, the limit is usually determined by the electronics employed for the light intensity measurements. Turbidity is even less of a problem than in fluorescence analysis, e.g. opaque polypropylene tubes may be used.

7.3.2 Key Parameters Guiding the Design of Luminometers

In essence, there are several key parameters in luminescence analysis:

1. Reaction rate
2. Volume of the reaction vessel
3. Efficiency of light collection
4. Sensitivity and noise level of the light detector

These parameters result in experimental arrangements (Figure 7-2) which in many respects differ from those in absorption spectrophotometry and fluorescence analysis. The unique importance of the reaction rate in determining the analytical response is discussed in Chapter 2. As explained above, choosing the volume of the luciferase solution determines the final concentration of the reagents and thus the efficiency of light production. In addition to this possibility of enhancing the emission intensity, small volumes have a more favorable geometry for light collection, but may require special sample handling. Attention should be paid to the fact that the entire sample volume participates in the light production, and preferably the global emission of the sample cuvette should be recorded. In conventional photometric methods, only the volume traversed by the incoming beam participates in the analytical response. In this case, the instrumental design is dependent on the fact that the beam has to be recorded in absorption spectrophotometry but kept from being recorded in fluorometry. There are some special demands on the light detector and the electronics for luminescence

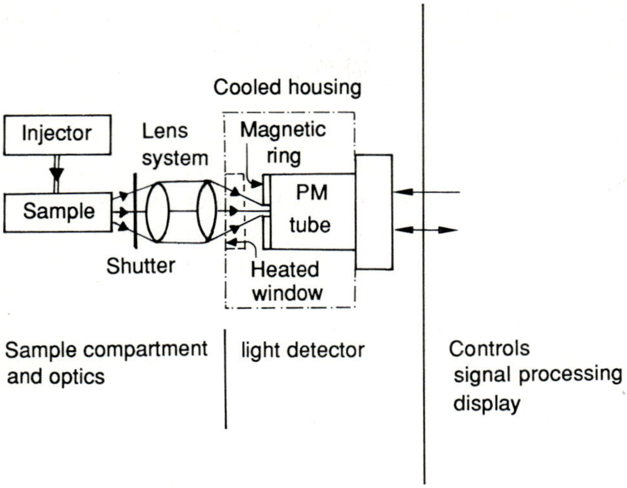

Figure 7-2. The main components of a sensitive luminometer.
The sample compartment is equipped for rapid mixing and joined to the optics for focusing the emitted light on the photocathode, which is protected from undue light exposure by a shutter. The light detector has a photomultiplier tube, which is connected to a pulse height discriminator. The sensitivity is enhanced by decreasing the electronic noise, which can be achieved by cooling the tube and shielding off the photocatode, except for a small central area. Fogging of the optical surfaces should be avoided when cooling. Connections for high voltage supply and controls are indicated.

measurements as compared with those employed in spectrophotometers and fluorometers. The light intensities are much lower and the intrinsic noise of the detector is a more severe problem. The light detector of a fluorometer may in principle also be used for luminescence analysis, but only in cases of exceptionally strong emission obtained with a high concentration of the substance to be determined.

7.4 Instrumental Facilities for Appropriate Recording

7.4.1 Sample Compartment and Light Collection

The versatility of a luminometer is highly dependent on the design of the sample compartment. If not to be used only for fixed routine analyses, acceptable standard equipment should be capable of accommodating various sample volumes preferably in the range beween 50 an 500 μL, allow for change from sample injection to manual mixing, and permit a choice between manual and mechanical feeding of the samples. In addition to these fundamental requirements, it may be desirable to have the sample compartment equipped with a thermostat and to be able to read the sample tubes from below. Far-reaching demands on the flexibility may call for an exchangeable housing, particularly when extended space is needed to allow for special analytical arrangements. This may, for instance, be necessary in order to permit flow systems or to provide for a sufficient temperature gradient when the reaction vessel is connected to a chilled photomultiplier by a light guide. It is indeed advisable to judge the ease of sample handling in relation to the construction of the sample compartment when the qualities of a luminometer are considered.

The use of the equipment for sample handling is essentially determined by the procedure chosen for mixing sample and reagent solution. Injecting the sample by means of a syringe may lead to a situation close to the ideal, with instantaneous creation of a homogeneous solution. In this case, mixing can be accomplished in a few milliseconds without inactivating the luciferase. With small samples, it is generally easier to inject the sample into the reagent solution than vice versa. A small buffer volume may be used to expel the sample, with an air bubble separating the two solutions. In case of automatic feeding of a series of samples, the reagent solution has to be injected instead, using a reloadable syringe without a driving buffer volume. Uniformity of injection is best ensured by a mechanically or air pressure-operated device, whereas electric arrangements are liable to disturb the light signal. In a setup for a preceding simple mixing of sample and light-producing solution, the use of vibrators can inactivate enzymes; while gentle tapping at the bottom of the tube with a fingernail or a similar mild procedure is harmless and generally quite sufficient. Care should be taken not to expose the light-producing solution to unnecessary illumination before or after mixing with the sample, which may induce disturbing background emission. Furthermore, there are tubes and reaction vessels which deliver a long-lasting disturbing emission after exposure to strong light.

When it comes to the question of a thermostat, it is in principle preferable from the point of sensitivity to operate at the temperature at which emission intensity is strongest. It is also suitable to work with a system which is relatively insensitive to temperature, in order to diminish the demands on the precision of the thermostat. It is important for the analytical processing that measurements of all samples and standards are performed at the same temperature. In practice, it is often most convenient to perform all processing and analyses at room temperature, but keeping the solutions on ice during the intermediate rest periods.

Reading sample tubes from below seems to improve the reliability when using immobilized enzymes. In some luminometers, this can be arranged with a mirror placed at an angle of 45° below the tube. Emission would be less impaired by internal reflection, if the bottom were spherical. As a consequence of the immobilization, the advantage of uniform light production throughout the solution is lost, and randomization of emitting foci may be desired; for instance, when a sedimenting enzyme suspension is used. However, small particles with a low density approaching that of the surrounding solution may help to maintain the stability of the suspension [9] and even permit reading of the emission from the side of the tube. Similarly, increasing solution density and viscosity by appropriate inert additives may improve suspension stability.

Arrangements for collecting a large proportion of emitted photons are generally desired, but the difficulties in effecting this are often underrated. A parabolic mirror large enough to keep the emitting volume aproximately in focus would greatly surpass the maximal aperture of current lens systems. However, the required size of the mirror makes this construction difficult. The prerequisites for collecting light emitted from solutions have been treated in detail by Malcolm and Stanley [10]. They drew attention to the considerable losses due to internal reflexions. Elevated signals after frosting the wall of a reaction tube may illustrate that light which strikes the wall of an untreated tube from the inside is reflected to a large extent, instead of reaching the light detector.

Light guides enhance the sensitivity considerably by permitting photon collection with a very high aperture [11]. An additional advantage is the fact that losses caused by internal reflexions are minimized, if the light guide is brought in good optical contact with the light emitter and the light detector by means of liquids with an appropriate refractive index. Designs for the application of light guides will be briefly discussed in the section on single cell assay, Chapter 9.

7.4.2 Solid-State and Vacuum Cells

Light detection in luminescence analysis usually needs time resolution, i.e. involves measurement of the time course of the emission. Spatial information on luminescence is only of interest in particular situations; for instance, when luminescence is used in an arrangement similar to autoradiography. Various photoelectric devices are used to follow the time course of the emission. In these devices, the stream of photons gives rise to an electric current, which is measured. There are two approaches to photoelectric detection, either solid-state photodiodes, or vacuum phototubes. With a vacuum tube, one may distiguish between a vacuum diode and a photomultiplier. The principle for creating the photo current is different in the solid-state and the vacuum cells. In the

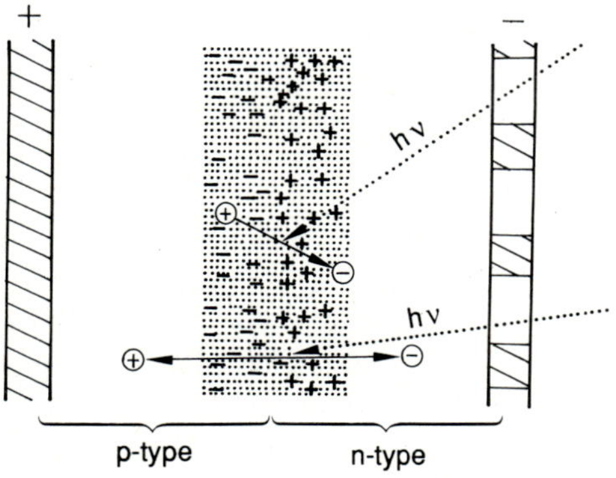

Figure 7-3. The principle of a solid state photodiode.
The diode is composed of two semiconductors of n- and p-type, joined together and to metal contacts, one of them shaped as a grid to permit the passage of light. The basic material of the semiconductor is usually silicon. In the n-type, conduction through electrons predominates, due to the addition of an element of high valency. Correspondingly, in the p-type material an element with low valency is added to provide electron vacancies (holes), and hence conduction through positive movable charges. A narrow p-n barrier (dotted area) arises at the junction, due to atoms depleted of their movable charges, creating an inner potential. A pair of movable charges, a hole and an electron, is created by an incoming absorbed photon and separated by the inner potential of the p-n barrier. The separated charges give rise to a photo current when collected on the metal contacts at the two sides. The drawing is not to scale, in reality the diode is a thin sandwich and consists typically of a 0.001 mm n-layer on top of a 0.5 mm p-layer.

solid-state photodiode, a charge pair, consisting of an electron and a positive hole, is created when a photon is absorbed in a photosensitive layer. The electron and the hole are separated by an inner potential barrier in the material, resulting in a photopotential (Figure 7-3). In the vacuum cell, electrons are emitted into the vacuum when photons strike a photosensitive surface, the photo cathode. The electrons are collected on a positively charged electrode, the anode, producing a photocurrent. The vacuum photodiode contains only two electrodes, the photocathode and the anode. The photomultiplier includes a series of extra electrodes, called dynodes, in the path of the electron flow (Figure 7-4). The electrons emitted by the cathode are attracted to the first

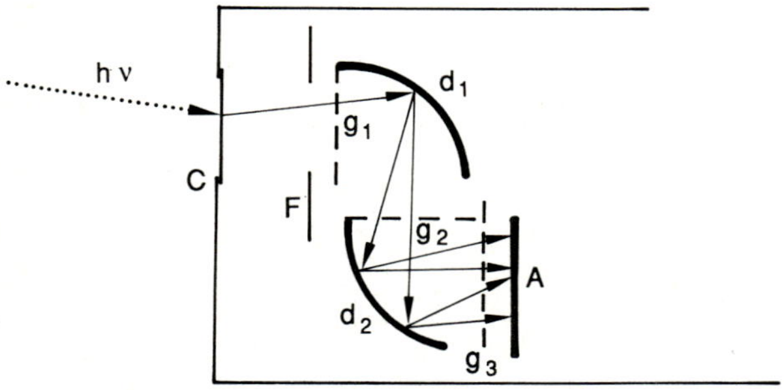

Figure 7-4. Illustration of the working principle of a photomultiplier tube.
To the left is shown how an incoming photon reaches the photocathode C, expelling an electron which is focused by the electrode F and accelerated by the grid g_1, of a dynode d_1. In this case, each incoming electron gives rise to an outflow of many more electrons, which are accelerated onto a second dynode by the grid g_2, which again causes an increase in the electron flow. The electrons from g_2 are accelerated by grid g_3 and collected on the anode A. For simplicity, a cell is shown with only two dynodes and with arrows indicating a doubling of the electron flow at each dynode. A real cell has many more dynodes and greater amplification at each dynode. The different electrodes are kept at successively increasing positive potentials, so that the overall voltage across the cell may be in the thousand-volt range.

dynode, which receives a suitable electric potential by an outside voltage source. The dynode has the property of emitting several electrons for each incoming electron. These emitted electrons are attracted to a second dynode with a somewhat higher positive potential, at which still more electrons are emitted, and so on. After a chain of dynodes, typically between 8 and 12, the electrons from the last dynode are collected on an anode and processed for display. A few incoming photons thus initiate a cascade, resulting in a large number of electrons, i.e. a considerable current amplification.

7.4.3 Quantum Efficiency and Sensitivity of the Light Detector

The efficiency of a conversion of light to electricity is usually given in terms of quantum efficiency. It is defined as the number of electrons given off for each incident photon. It is related to the energy of the photon, i.e. the wavelength of the light, and should be specified in relation to the spectrum of the light to be measured. The solid-state cell has the highest quantum efficiency and may in extreme cases with monochromatic light, suited for maximum sensitivity, approach unity. The quantum efficiency of vacuum cells is on the whole considerably lower. However, solid-state cells have a much lower inner resistance than the vacuum devices, resulting in a high background level and difficulties in measuring low light intensities. The signal to be measured is "drowned" in the noise of the dark current. The situation is much better for a vacuum cell, particularly with a photomultiplier. As a rule of thumb, a vacuum photodiode may measure intensities which are 100 times weaker than those detected with solid-state cells. With photomultipliers, there is at least another 10 fold increase in sensitivity, i.e., a good photomultiplier is typically more than 1000 times as sensitive as a solid-state cell.

The solid-state cell, usually a silicon photodiode, is a very rugged construction, which therefore lends itself to use in mobile constructions in the field. Despite its need for higher light intensities than the vacuum cells, its sensitivity range meets the demands of several applications. When coming down to the ultimate sensitivity levels of vacuum cells, there is often a need for special sample handling arrangements, which do not lend themselves to field use. This parallels the higher fragility of the vacuum cells, which calls for special caution during operation, particularly in the case of photomultipliers.

Photomultipliers require a high voltage source for the dynode chain, equipped with a high degree of stability, as small voltage changes result in considerable changes of the output. There is also a marked susceptibility to temperature changes. To obtain maximal gain and a low dark current, one should work with chilled multiplier tubes. Their dark current level is reduced approximately by half, if the temperature drops by 10 °C. The main effect is obtained already just below −10 °C and a good piece of equipment is a Peltier element, in which outside heat production is dissipated by cooling water. The multiplier tubes show high degrees of individuality, and their performances depend on their history. It takes days for a tube to recover from exposure to intensive light. Accordingly, the importance of a well-functioning shutter and diaphragm should be recognized. It is important that the shutter does not introduce undue disturbances, e.g. from electric or magnetic fields. A safe solution may be a pneumatically operated shutter. Recovery from exposure to high voltages is comparatively fast but may also pose a problem during measurements. Relaxation to stable temperature characteristics is slow. Overexposure to heat may cause permanent damage to cells. Magnetic fields from outside may influence the signal. Such fields may arise from auxiliary equipment, but can to a great extent be neutralized by a shield of mumetal.

The photomultiplier has to be operated with an amplifier. This is primarily for current amplification, so that the signal may be displayed, e.g. on a recorder or an oscilloscope. Rise times of the tubes may be very fast, often in the nanosecond range, and any problem in following fast intensity changes of a luminescence source is generally due to limitations of the auxiliary equipment. Linearity between light intensity and output is maintained over a wide range typically exceeding 10^6. However, at the high sensitivity end, care should be taken not to expose the tube to an unduly high voltage, as this may damage the tube temporarily or even permanently. At the low sensitivity end, the application of too low a voltage to the dynode chain results in a loss of linearity; and with high light intensities, it is necessary to see to it that the incoming light is weakened by appropriate means, e.g. by adding an optical filter or reducing the aperture.

7.4.4 Amplification of Light Detector Signals

There are basically two means of processing the output from the light detector: either the current is measured or the number of pulses recorded. Current measurements are applicable to all type of detectors; whereas pulse counting in practice solely applies to photomultipliers.

An arrangement using direct current measurements is displayed in Figure 7-5. The current of the photodetector is fed into a load resistor R. The value of R determines the

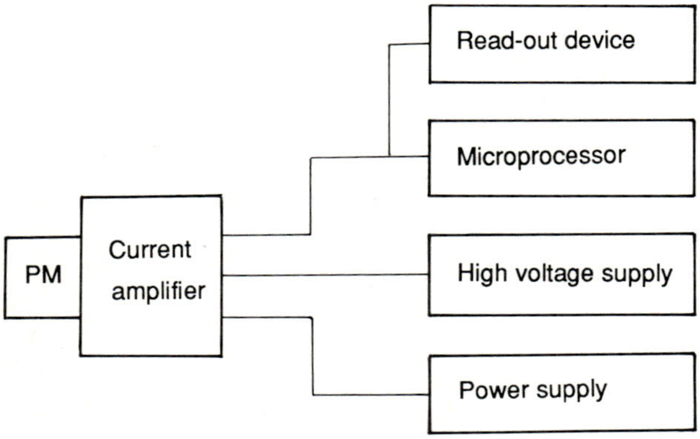

Figure 7-5. Block diagram of a common system used for light detection and signal processing. The output from a photomultiplier PM is fed into a current amplifier, which is provided with controls for adjusting a feed-back resistor and capacitor. The high voltage for the photomultiplier and the power for the amplifier are provided by separate units (the boxes on the lower right). The output from the current amplifier may be used directly to drive a read-out device, such as a tracking voltmeter, a recorder, or an oscilloscope; or it may be processed further, e.g., to provide information about rate constants or about the integral of emitted light.

output voltage and thus the gain. The output of the photodetector is usually not powerful enough to drive the display equipment directly. For connecting a recorder or an oscilloscope, an intermediate current amplification is needed, today a piece of standard equipment, adding very little to the overall noise. The outcoming signal is then supplied over a low resistance, permitting connection to various types of display devices. The time constant τ is determined by R and a capacitor C, $\tau = RC$. It is preferable to work with equipment which permits R and C to be selected quite independently. For weak emissions and rapidly occuring events, a compromise in the setting of R and C may have to be reached for sufficient sensitivity and time resolution. Only photomultipliers are capable of delivering a suffiently high burst of electrons for each primary photo electron given off at the photocathode, i.e. the millionfold amplification in the dynode chain yields pulses of electrons, which can be distinguished from the background noise.

The block diagram of Figure 7-6 exemplifies a photon-counting equipment. This method has an advantage, if the light levels are very low. The light absorbed by the photocathode gives rise to electron pulses which are fairly uniform in size; whereas the noise, which should be subtracted, is less uniform. It is possible to select and count only pulses which have a certain intensity. To a large extent, noise is produced by random

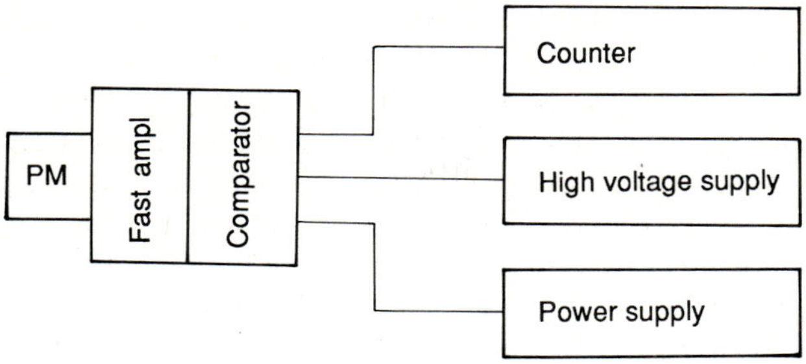

Figure 7-6. Block diagram of the main components of a detector system used for photon counting.
The pulses from the photomultiplier PM (powered by the high voltage supply) are amplified without distorsion by a fast amplifier. The comparator is used to sort out inappropriate pulses, discriminating those which are too strong or too weak. The extent of discrimination and the power is controlled by the power supply unit.

emissions of various parts of the dynode chain, resulting in pulses weaker than those magnified all the way from the cathode. There are also some very large pulses to be discriminated, arising from incoming high energy rays, e.g. cosmic radiation. The discriminator setting for the exclusion of false weak and strong pulses also leads to loss of a few true pulses; but these are only very few compared to the number of recorded counts.

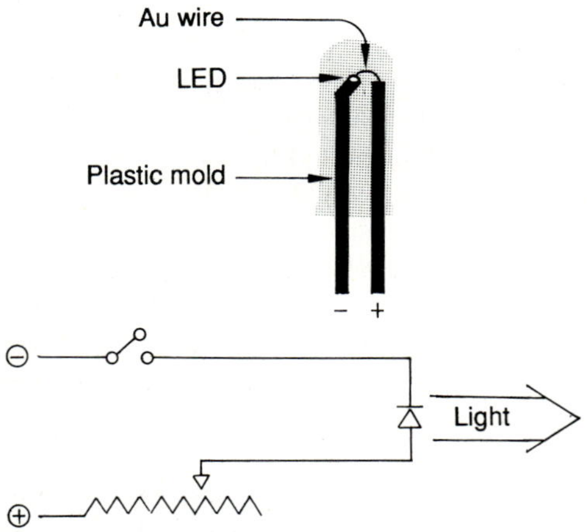

Figure 7-7. Typical appearance of a light-emitting diode (LED), which is an excellent light source in connection with adjusting controls and standards in luminescence measurements. The light-producing mechanism is essentially the reverse of that experienced in a photodiode (cf. Figure 7-3). Charge carriers, i.e. conduction band electrons and valence band holes are produced by application of voltage. Radiation is emitted by the recombination of the charge carriers in the semiconductor junction. Frequently used LED materials are GaAs and GaP. The overall size may be a few mm.
The electric circuit for operating the cell is very simple. A stable voltage of about 5 V is supplied by an electrochemical cell and the output controlled by a stepwise variable resistor in the range 300 kΩ to 10 000 MΩ. When checking light pulses of short duration, the switch should be equipped with a suitable timer.

It is generally desirable to preset the equipment with adequate adjustment of all controls, before a series of analyses is carried out. This is particularly the case with photomultipliers, which should be operated under very stable conditions. Thus, a change in high voltage setting should be avoided during measurements, and a long time should be allowed for temperature equilibration with the settings selected. For adjusting the controls and checking the equipment, the use of an easily operated light generator is of great help. We have found that a light-emitting diode is excellent, both for applying current measurements and for photon counting (Figure 7-7). The diode can be used to check appropriate sensitivity, linearity of output, adequate dampening without distortion of the time profile, and the noise level while equilibrating the tube. The emission characteristics of the diode can be defined by a simple circuit powered by a battery (Figure 7-7). The diode also has the advantage of emitting the same light spectrum, independently of the intensity. The use of standards of slowly decaying radioactive compounds in a scintillating fluid is often suggested as a suitable light source for similar purposes. We have made the experience that such light sources have considerable disadvantages, particularly in relation to calibrating the photon counting equipment, where the irregular shower of photons may lead to inappropriate settings of the pulse discriminator.

7.4.5 Commercial Instruments

The demands on a luminometer are generally connected with its ability to measure light at low intensities, its versatility in sample handling, and convenience in signal processing and display. How this happens in the individual commercially available instruments is described elsewhere, together with information about construction and mode of operation [12–14]. A comparison and evaluation of the different instruments is outside the scope of this presentation. However, some general guidelines may be proposed when selecting an instrument to satisfy the actual analytical demands.

A basic question is whether the sample compartment is likely to be adequate. It should be realized that its versatility in the final processing and handling of samples is directly linked to its design. Since this matter has been treated in detail above (Section 7.4.1), it may be sufficient in this context to point out that space should be allowed for possible modifications, in case the sample housing is fixed and not easily exchangeable.

Furthermore, it should be kept in mind that the photosensor and its auxiliaries are the heart of the instrument. As illustrated in Figure 6-3, it is often preferable to work with diluted samples and thus weakened signals, in order to reduce influences from disturbing reactions. This implies that the sensitivity must in all cases be appropriate and attention be paid to the signal-to-noise ratio. With more advanced instruments, the means used for noise suppression should be recognized, such as cooling of the sensor and electric and magnetic shielding. It is, of course, also necessary to ensure that electronics and sensor are properly matched. The time constant of the equipment is an important factor in this context, particularly when it is to be used in connection with flash recordings or the delayed appearance of a maximum, cf. Table 7-1. Instruments which permit the separate adjustment of feed-back resistor and capacitor may offer advantages.

The way to arrange for the display and processing of signals should be carefully chosen. With many instruments, these means are quite limited. The output may, for instance, only be available as a set of digits, sometimes in relation to a fixed recording time and without possibilities of following the time course of the emission. This may be adequate for a very limited type of application, but is seldom satisfactory in a research laboratory. Instead, access to separate signal outlets should be looked for, permitting modifications in signal processing and display, if so desired. Linkage to microprocessors of various kinds are popular with many of today's instruments and may yield useful information about rate constants, integrals and such like; but it is important to realize that these instruments are often built with a particular analytical use in mind, thereby limiting possibilites for other uses. Coupling to an available ordinary PC may instead provide all necessary information [15].

As already discussed in Section 7.4.3 the level of detectability, i.e. the minimum amount of analyte which can be measured, is mainly determined by the photosensor. The photocurrent from a solid-state diode can be sufficiently amplified to allow the analysis of small samples, but vacuum cells are preferred in most laboratories. However, in special applications in which ruggedness is important, e.g. for outdoor and mobile use, a photodiode may be preferable. Current measurements, generally carried out by reading the voltage across the load resistor, permit analyses at or just below the pmol level. Photon counters equipped with adequate outfit for pulse height discrimination are required for work at lower levels. Still, in most applications, there is little to be gained by one method over the other. Financial factors are often decisive, and more sensitivity instruments generally cost more. Photon-counting equipment tends to be more expensive than instruments that measure direct current, even with instruments of the same sensitivy.

References

[1] Lee Y, Jablonski E, DeLuca M. Anal Biochem. 1977; 80:496–501.

[2] Ugarova NN, Brovko LY, Beljajeva EI. Enzyme Microb Technol. 1983; 5:60–64.

[3] DeLuca M. In: Kricka LJ, Stanley PE, Thorpe GHG, Whitehead TP, eds. Analytical Applications of Bioluminescence and Chemiluminescence. New York: Academic Press 1984: 111–123.

[4] Jablonski E, DeLuca M. Proc Natl Acad Sci USA 1976; 73:3848–3851.

[5] Roda A, Girotti S, Ghini S, Carrea G. J Biolum Chemilum. 1989; 4:423–435.

[6] Ågren A, Brolin SE, Hjertén S. Biochim Biophys Acta 1977; 500:103–108.

[7] Brolin SE, Ågren A, Wersäll JP, Hjertén S. In: Schram E, Stanley P, eds. Analytical Applications of Bioluminescence and Chemiluminescence Westlake Village, California; 1979: 109–121.

[8] Brolin SE, Hjertén S. Mol Cell Biochem. 1977; 17:61–73.

[9] Brolin SE, Ågren A, Ekman B, Sjöholm I. Anal Biochem. 1977; 78:577–581.

[10] Malcolm PJ, Stanley PE. Int J Appl Radiat Isot. 1976; 27:397–430.

[11] Brolin SE, Berggren P-O, Naeser P. In: Kricka LJ, Stanley PE, Thorpe GHG, Whitehead TP, eds. Analytical Applications of Bioluminescence and Chemiluminescence. New York: Academic Press; 1984: 479–482.

[12] Van Dyke K. In: Van Dyke K, ed. Bioluminescence and Chemiluminescence: Instruments and Applications, Vol 1. Florida: CRC Press; 1985: 83–128.

[13] Stanley PE. In: Kricka J, Carter TJN, eds. Clinical and Biochemical Luminescence. New York: Marcel Dekker; 1982: 219–260.

[14] Stanley PE. In: DeLuca MA, McElroy WD, eds. Methods in Enzymology New York: Academic Press; 1987: 587–603. (Bioluminescence and Chemiluminescence, Part B; vol 133).

[15] Ågren A, Brolin SE, Lindfors M. In: DeLuca MA, McElroy WD, eds. Bioluminescence and Chemiluminescence Basic Chemistry and Analytical Applications. New York: Academic Press; 1981: 575–581.

8 Fields of Application

8.1 A Biological Outlook on Analytical Applicability

In bioluminescent species, the production of light represents part of the cellular metabolism. This provides ample opportunity for coupling to various reactions when designing analytical reaction schemes in vitro. It is particularly common to use it for the measurement of ATP as well as ATP-producing metabolites with the firefly system. There is also a manifold of analyses that are possible by coupling the bacterial luciferase reactions to conversions between $NAD(P)^+$ and NAD(P)H. As pointed out in Chapter 2, the analyses mimic metabolic pathways; and their advantage is the individuality of the concerned enzymes, resulting in a high degree of specificity of the overall analytical reaction. The high specificity made possible by bioluminescence analysis is a definite advantage and may be compared to the same quality of the otherwise different immunochemical reactions. Combining immunochemistry and bioluminescence analysis has proved useful and has led to a new rapidly developing technique [1, 2], bioluminescence immunoassay (BIA).

Today, there are cases in which the application of several native bioluminescence systems is hampered by limited access to biological material. It may be expected that in the future gene technology will be able to provide sufficient supply. Cloning and expressions of the firefly luciferase gene has been carried out in cultured mammalian cells and bacteria [3, 4]. Thus, it is evident that production of this enzyme should become independent of the breeding of the fly. The scarcity of the earthworm *Diplocardia longa* may be taken as an example of a case in which limited amount of material has prevented analytical application. This species contains a luciferase capable of catalyzing light formation from hydrogen peroxide much more specifically [5, 6] than the reaction with luminol, which is nowadays generally employed in studies of the respiratory burst (cf. Section 1.4.1). The *Diplocardia* luciferin-luciferase system could be a useful tool in research on free oxygen radicals and scavengers, but the large enzyme ($M = 300\,000$) makes cloning difficult. One can hope that the large number of potential applications could be strong incentive for work in this area. The clam *Pholas dactylus* should likewise be of interest for studies of oxidative processes, but it is threatened by extermination. Again, these are large molecules, with $M = 310\,000$ for luciferase and $M = 34\,600$ for luciferin, in this case a photoprotein [7].

Besides the close relationship with important biochemical processes, bioluminescence reactions have potentials because of their wide concentration range and the possibility of in situ analysis in biological specimens. Visualization and evaluation by image intensification is an attractive line of development. There are parallels to the development of fluorescence microscopy, which initially yielded valuable qualitative information and was then followed by facilities for quantification. An example is the light emission initiated by calcium ions after the deposition of aequorin in a cell, as referred to in the section on single cell assay, Chapter 9. The production of luciferase in the cell is an efficient way of visualizing the cells. A luminescence assay may then be carried out in vivo by subjecting the cells to an appropriate environment of reactants. The disadvantage of the normal analytical procedure, which makes it necessary to destroy the cell and to rely on in vitro analysis, is thus circumvented. This direct visualization has become a valuable tool in recombinant DNA technology involving the cloning of genes that lead to the production of luciferase.

We can conclude that there are two different kinds of applications. On the one hand, there are true analyses that often focus on collecting biochemical information under in vivo conditions. On the other hand, it is a matter of being able to monitor a system either with respect to mapping the emission in relation to the fine structure of a specimen, or following the dynamics of an event.

8.2 Biomass Determinations in Microbiology and Cell Research

8.2.1 Means of Quantification

Biomass is a concept derived from trying to quantify the amount of living matter. The quantifications may in some cases be fairly straightforward, but in other cases it is more ambiguous. A conventional approach is to count the number of living cells; this is, for instance, useful for cell cultures. The amount of genetic material DNA, being the same for every cell of a certain species not engaged in division, may serve as an indicator of biomass. Biomass may also be quantified through the amount of accumulated product, such as protein or organic carbon, and is in this case based on time and producing ability. This ability is reflected in the amount of ATP in the cell, which became conveniently determinable by the introduction of bioluminescence analysis (cf. Section 1.4.2). Estimates of biomass based on ATP measurements emanate from microbiology [8-12]. Particularly when it comes to unicellular cultures, the quantity of ATP has been found to correlate well with other measures of the cellular amount.

8.2.2 Use of ATP

The use of ATP as a measure of biomass is based on its presence in all living cells. To be able to exist, all of them depend on a number of energy requiring processes. Viability therefore sets a lower limit to the required amount of ATP. Hence, variations in the concentration of ATP cannot sink below this limit, which also means that the presence of living cells is generally indicated by the presence of ATP. The activities of certain cells, such as mobility, secretion, or maintaining concentration gradients, require ample energy. The upper limit of the ATP concentration may thus be expected to vary considerably. Although it should be kept in mind that the energy supply may depend more on high conversion rates than on the concentration of adenylates, measurement of ATP is a useful approximation of the number of cells present.

Pollution of water or contamination of food may easily be shown by measurements of ATP. The possible quantification is generally sufficient to decide whether water is potable, suitable for bathing, etc., or whether food is eatable [11-13]. Another application is the detection of bacteria in normally sterile body fluids. In particular, techniques have been developed for diagnostic use with respect to bacteriuria [12, 14-18]. In such analyses, it is important to eliminate interference by free ATP in solution and in suspended human cells. This is accomplished by adding a small amount of apyrase to the sample and by treating with a mild detergent, which destroys the human cells but leaves the bacteria intact. In a subsequent step, the more resistant membranes of the bacteria are destroyed and their ATP is released and analyzed. In exceptional cases, the presence of mitochondria with resistant membranes may be a source of error, if they are lysed together with the bacteria.

In the aforementioned situations, the ATP determinations serve the purpose of demonstrating the presence and quantity of living cells in various environments. A similar application is following the changes of biomass during microbial growth experiments. Samples taken at different times yield information on the growth rate. As the same type of sample is used in the different analyses, errors arising from variations in the proportionality between number of cells and analytical response are largely eliminated. With these techniques it is therefore possible to determine growth curves with sufficient precision. Increase in ATP can be employed for the classification of microorganisms by selecting a culture medium which fits the particular demands of the strain to be isolated [10]. It is possible to monitor increased microbial growth as a result of stimulation by various agents, e.g., vitamins [10]. Likewise, it is possible to quantify retarded growth in the evaluation of antimicrobial agents, such as antibiotics and antifungal drugs [19-20]. In these tests, leakage from damaged cells deserves attention. The interference of such ATP may be eliminated through treatment with apyrase. It is also possible to collect further information through monitoring this outflow. Growth retardation can, moreover, be used to demonstrate antibodies in body fluids. This requires access to the bacteria in question, which may, for instance, be obtained from a pure bacterial culture.

It is possible to extend quantification based on bioluminescence to cells and tissues of human or other mammalian origin [21, 22]. It should be recalled, however, that ATP measurement reflects the cellular content better than many other method of estimating biomass. Extracellular products, such as connective tissue, may play a predominant role when protein content, dry weight, or organic carbon are used for biomass determination, see Section 8.2.5.

8.2.3 The Adenylate Pool (ATP+ADP+AMP)

The fast turnover of adenylates raises questions as to the reliability of ATP measurements for quantifying biomass. In many organisms, a 24-hour adenylate turnover results in the formation of quantities of ATP as large as the entire weight of the organism. It can be expected that equilibria within the adenylate pool are subject to change during this interconversion. Such changes have, in fact, been observed during the analysis of cell specimens under different metabolic conditions (Figure 8-1). The changes of the equilibria during handling and preparation of specimens for analysis are of particular importance for the reliability of the analytical measurements. A substantial and rapid ATP decrease may occur, unless particular precautions have been

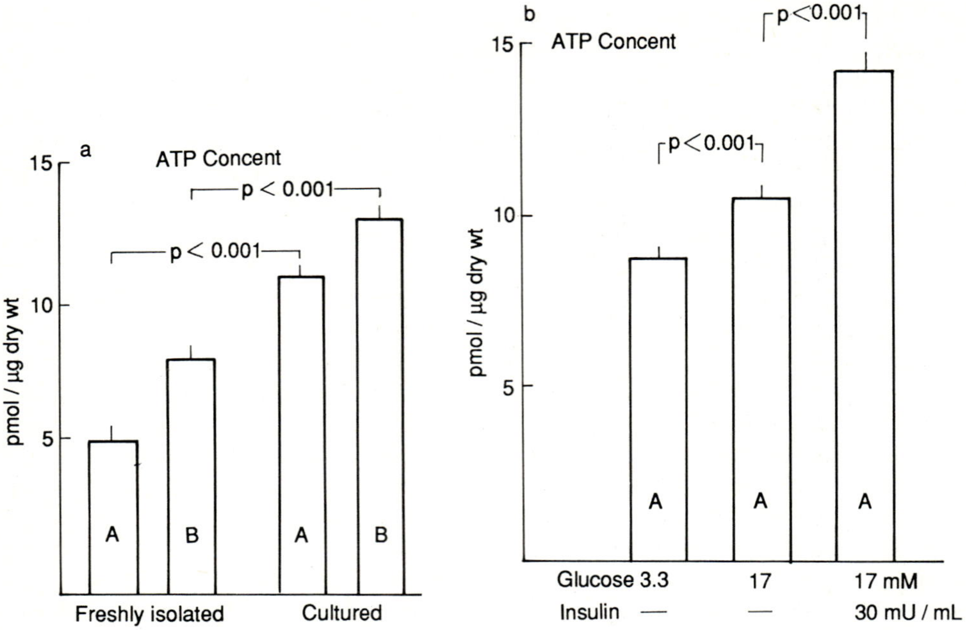

Figure 8-1. ATP content in pancreatic islets from guinea pigs, in cells producing glucagon (A) and in cells producing insulin (B).
a. Demonstration of recovery of the islets by culture.
b. Vitality in culture, increased by glucose and insulin stimulation. It can be seen that ATP is a rather uncertain measure of biomass, unless the metabolic conditions are properly specified. Redrawn from [54] and [55].

taken during sample processing. This is illustrated by actual measurements performed on the decay (Figures 8-2, 8-3 and 9-8).

Thus, ATP concentrations are inclined to change during preparation, if this is not prevented by special precautions. In the analysis of tissue specimens, quick-freezing and lyophilization are routine procedures which work well. A similar preparative procedure can also be used for cell suspensions, by starting the preparations with centrifugation, forming a pelleted sample comparable to a tissue specimen. However, significant changes in ATP concentration may already occur during this process. Alternatively, it is possible to add the cell to an extractant, which in case of tissue specimens must be preceded by a homogenization step. Again, considerable changes in the ATP content may occur. In some cases it has been possible to compare different extraction methods; in such cases, the true value of intracellular ATP is often set equal to the result obtained by the method which leads to a maximum yield of ATP. A comprehensive survey of possibilities and problems of extraction has recently been presented [23].

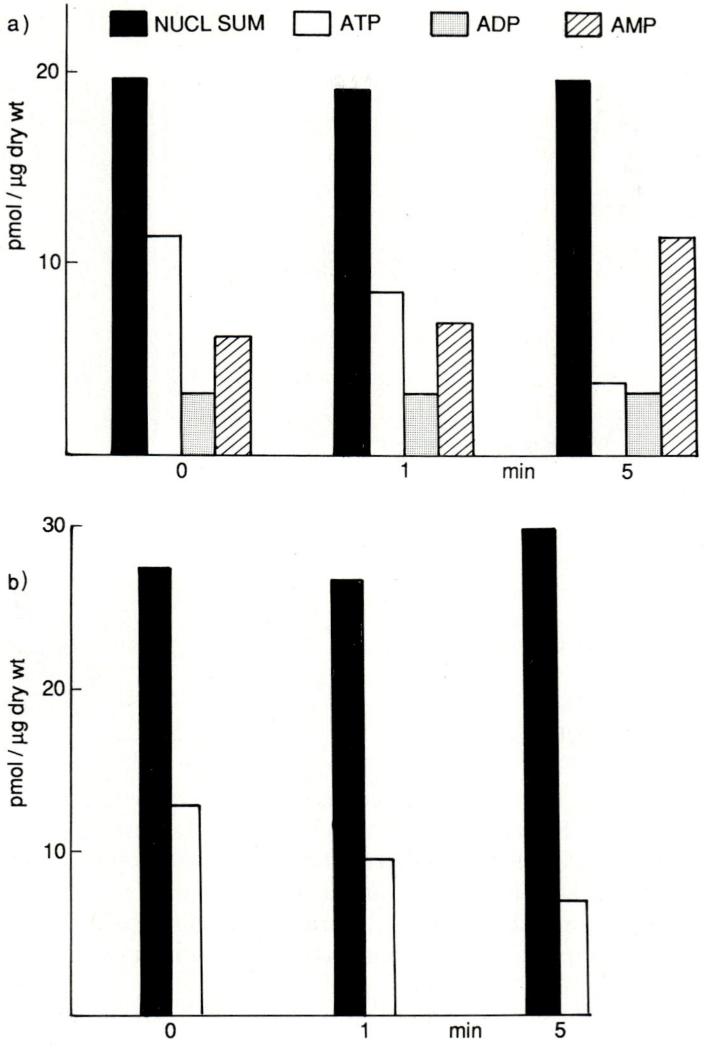

Figure 8-2. Effect of hypoxia on the content of adenine nucleotides in insulin-producing islet cells. Redrawn from [56] and [54].
a: Adenylates in islet cells isolated from mouse pancreas with the vascular supply interrupted during the times indicated. After rapid freezing, islet specimens are dissected from lyophilized sections. The decrease in ATP concentration with hypoxia is not accompanied by a detectable alteration of the adenylate sum.
b: The adenylate sum (filled bars) and ATP contents in cultured islets from guinea pigs. Again, it can be seen that the adenylate sum stays essentially constant; whereas the ATP content decreases markedly when the oxygen supply is interrupted. The ATP level can be restored by transfer to culture in an oxygenated medium.

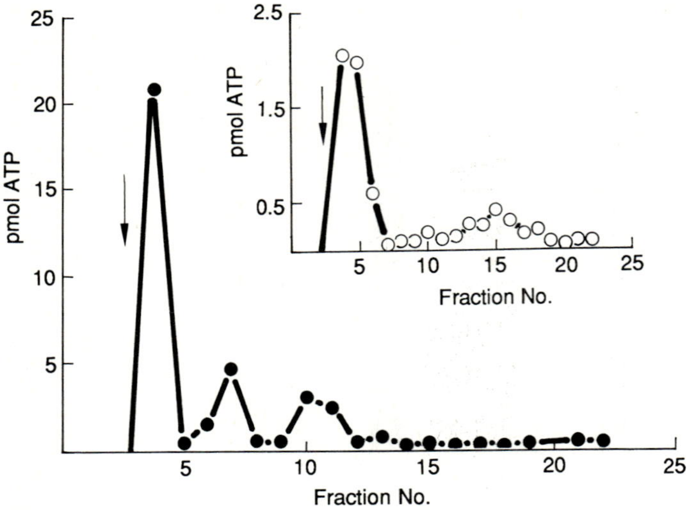

Figure 8-3. ATP extracted from rat ureters using laminar elution with saponin [57], (cf. Section 9.5).
Main figure: Ureter cannulated in a living animal.
Insert: The second ureter 3 hours after killing the animal.
It can be seen that the ATP level is drastically changed.

8.2.4 The Concept of Biomass

The concept of biomass is not clear. The prefix "bio" points to analyses of living cells. In necrobiotic cells, the amount of ATP decreases rapidly, and such cells become more and less excluded from a biomass estimate as determined by the ATP value. Although ATP analyses are convenient and suitable for monitoring changes in a cell population, evaluations in terms of biomass may be uncertain, particularly in the presence of metabolic disorder, since the number of cells which can recover may be underrated, due to their lower ATP content.

A pool of adenosine phosphates is a prerequisite for the energy conversion processes. A large amount of adenylates is a potential for a high energy turnover. The sum of the adenylates is not affected by interconversions between ATP, ADP and AMP and should therefore be a more reliable measure of biomass than ATP alone (cf. Figure 8-2). For practical purposes, it is necessary that the analysis can be performed as a one-step procedure. It is expected that appropriate reagents for this soon will be easily available.

Bioluminescence offers alternatives for estimating biomass, apart form adenylate analyses. In particular, an assay of the sum of FMN and FAD may be attractive and may develop into an easy routine method. FMN and FAD are important compounds in the production of energy by oxidative phosphorylation. During acid extraction, FAD is easily hydrolyzed, which means that the two nucleotides may be obtained together. With a simplified bacterial luciferase setup, the amount of FMN may be analyzed through electrochemical reduction to $FMNH_2$ at a platinum electrode [24] (Figure 8-4). The versatility of this analytical procedure is quite obvious, but a closer evaluation of its

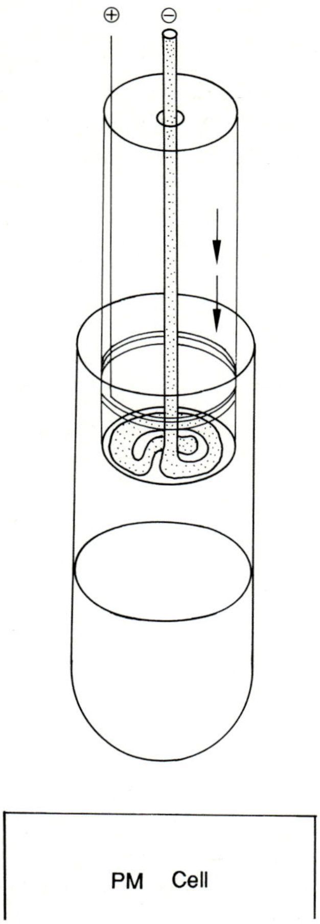

Figure 8-4. A simple arrangement for light production by electrochemical reduction of FMN in a bacterial luciferase solution.
A platinum rod (1 mm) is bent and flattened at its end and then mounted in a glass tube with thick walls. The flattened end serving as a cathode is easily polished to maintain reproducibility. A slender platinum thread tied round the glass tube forms the anode. (With oxidized platinum, this setup is not suitable for testing the sensitivity and reproducibility.)

applicability is required. An advantage may be lower leakage of flavines, compared to cell adenylates, as the flavines are more tightly incorporated into the subcellular structure. Another point is the financial advantage of these chemicals over those needed in the firefly assay. Less sensitivity may in some situations be a disadvantage of flavine assay.

Using ATP, the sum of adenylates or flavine nucleotides as indicators of biomass implies that the concept of biomass is thought of as a measure of energy producing ability. There is, however, a marked difference between using ATP and the adenylate sum or flavine nucleotides. ATP is very sensitive to the time factor, and only traces are present in dead specimens left at room temperature for a short time; whereas the adenylate and flavine pool is more unaffected.

8.2.5 Various Methods Compared

Several methods of quantifying biomass are applicable. However, the result will vary with the method and the type of specimen being investigated. This is illustrated in Figure 8-5, where the results of different methods are compared for different analytical

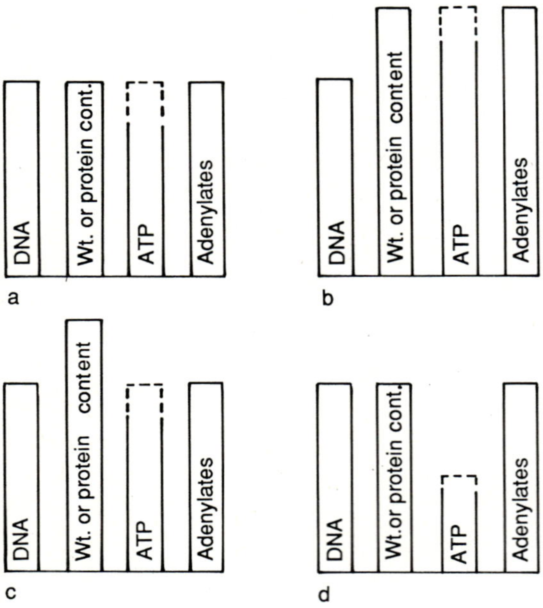

Figure 8-5. Illustration of differences obtained with various estimates of biomass. The samples are assumed to contain the same number of cells from the same species. a: Sample composed of uniform cells of a reference size. b: Sample composed of same number of cells, but larger than in a. The amount of DNA remains unchanged, whereas weight, protein and adenylate content increase. c: Sample as in a, but also containing tissue components. Weight and protein content are increased, whereas the other quantities remain unaltered. d: Sample as in a, but suffering from a metabolic disorder. ATP decreases, but the other parameters stay essentially unaltered. Dotted lines are used to indicate variations in ATP values, depending on the cell conditions during sampling.

situations. The methods compared are determination of 1. DNA; 2. weight or protein content; 3. ATP; and 4. the adenylate sum. It is assumed that the number of cells and, accordingly, the amount of DNA stays constant in all samples. Larger cells are heavier and contain more protein as well as ATP and other adenylates. This implies that biomass methods must be used with caution when applied to a determination of the true cell number. Proper calibration may be a key to the problem. In many situations the size distribution of the cells remains the same in the different samples and relative changes of biomass are easily followed. The presence of tissue components other than cells does not affect the DNA value nor the amount of ATP or the adenylate sum, but brings about an increase in weight and often also of the protein content. Assay of ATP or the adenylate sum can thus be employed for a quantification of the cellular moiety. In fact, they yield results which are more proportional to the amount of metabolic active cytoplasm than to the cell number. It may be thought of as containing a correction for differences in cell size.

The problem of stability during the analytical preparation of the quantity to be analyzed is essentially limited to ATP, which may disappear rapidly, as pointed out above. However, even if the preparative errors in the assay of ATP can be more or less eliminated, we have to face the fact that the ATP concentration may be dependent on the metabolic state of the sample, as exemplified for mammalian cells. In microbial cells, the concentration of ATP may likewise vary and be influenced by such factors as oxygen tension and the growth phase at sampling [22, 25]. The uncertainty in the determination of ATP is indicated in Figure 8-5 by dots on the tops of the ATP bars.

The applications of the various methods for biomass determination may be summed up as follows. Cells are the basis of life, and the number of cells can be determined through their specific components, such as DNA, adenine nucleotides, and flavine nucleotides. Assay of ATP alone has been found to be a convenient measure for screening biomass in microbiology. In such applications, there is more concern about sufficient reproducibility rather than the assessment of a true concentration.

The assay of ATP in combination with other methods of determining biomass, e.g. the adenylate sum, is of particular value in screening for cellular damage, since necrobiosis is accompanied by a rapid decrease in ATP concentration, whereas the adenylate sum is more stable. When it comes to mammalian cells or tissue samples, we are not prepared to recommend the analysis of ATP alone for determining biomass, but we feel that the adenylate sum could serve as a quantification of minute samples and a measure of the cellular components of tissues.

8.3 Clinical Chemistry

8.3.1 Applications in Established Routines

The routines of clinical chemistry have been subject to a tremendous change over the last decades. The introduction of semiautomatic or automatic methods has replaced laborious test tube procedures. This means investment into heavy equipment of various sorts and the establishment of proven technologies for various analyses. It is unlikely that bioluminescence methods will replace well-established methods, unless profound advantages are evident. One such case may arise, if analysis becomes an important aid for early diagnosis and if this is of great importance for the clinical treatment of the patient. Appearance of creatine phosphokinase in the blood is a very early sign of cardial infarction and may conveniently be demonstrated by bioluminescence [26-28]. Other examples are brain damage caused by head injury, neurological diseases, or transient hypoxia, where the presence of adenylate kinase in the cerebrospinal fluid is difficult to prove by other methods of analysis [29].

8.3.2 Immunochemistry

An important area of application of bioluminescence concerns attempts to find a substitution for the radioactive label in Radio Immuno Assay (RIA). RIA is widely used in biochemical and clinical laboratories for the quantification of a large number of biologically important compounds. The method is based on a competition between specific antibodies and antigens. A common approach involves labelling one of two antigens and keeping an antibody immobilized. At increasing concentration of the other antigen, i.e. the compound to be assayed, there will be less of the label attached to the antibody and more of it present in solution. Other designs may be employed, but a common feature is that after a separation step, the assay is terminated by measurements of radioactivity. Since RIA is impaired by a limited half-life of isotopically labelled antigens or antibodies, psychological dislike of handling radioactive material, and inconvenient regulations for waste disposal, radiolabels have been replaced by fluorescent dyes and chemiluminescent compounds or by nucleotides and enzymes which can be safely analyzed by non-radioactive methods. Horse radish peroxidase has been frequently used as a label. In this case, the final chemiluminescence reaction is accomplished by letting the peroxidase catalyze the oxidation of luminol with hydrogen peroxide. Luciferin from the clam *Pholas Dactylus* has been reported to be more suitable than luminol, due to a higher light output [7]. There seems to be a great deal of competition between immunoassay methods, among which many are closely related to bioluminescence. Nucleotide labels and particularly ATP conjugates are potentially very useful [30], but their future use is as yet difficult to predict. It is very likely that improvements are possible with labels that are related to bioluminescence and to take advantage of the specificity of enzyme reactions. Pyruvate kinase has been covalently coupled to antibody molecules for bioluminescence determination with firefly luciferase, utilizing phosphorylation of ADP to ATP by phosphoenolpyruvate [31]. Glucose-6-phosphate dehydrogenase conjugates have been measured after addition of glucose-6-phosphate by catalyzing the conversion of $NADP^+$ to NADPH which, in turn, produces light in the bacterial luciferase reaction [2]. A bacterial luciferase conjugate has likewise been employed and measured by light production [32]. It is also possible to use aequorin as a sensitive and convenient label [33]. There is therefore ample evidence of various means of labelling in connection with bioluminescence assay.

8.3.3 Biopsies

The main work load in clinical chemistry concerns analyses of body fluids; but analyses of other samples, such as biopsies, are of growing importance. For the clinical application of bioluminescence analyses, the development of methods to obtain relevant samples is crucial. This pertains in particular to the biopsy-taking technique. The demands on the technique may depend on the kind of tissue to be sampled, but also on the access routes. For instance, biopsies from skin [34-39] and from mucous membranes [40, 41] may in this context be easily accessible sites. Kidney biopsies may, on the other hand, represent a situation where access to the organ presents a greater difficulty. An assembly of instruments are shown in Figure 8-6, all designed to allow

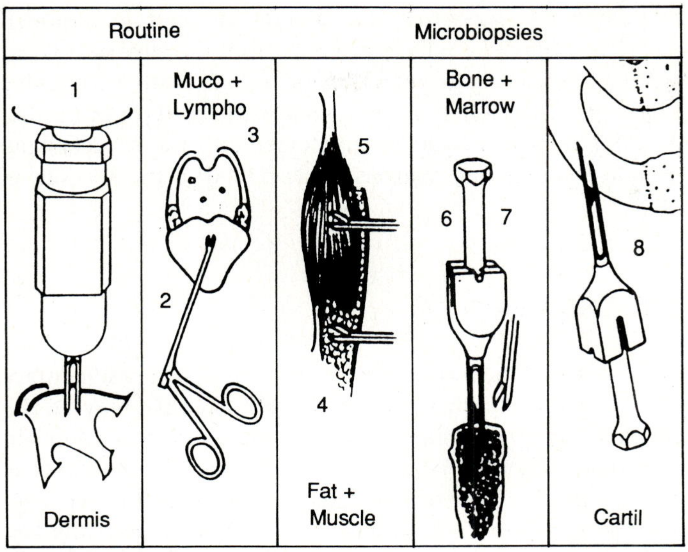

Figure 8-6. Instruments and methods for ambulant biopsies.
1. Powered micro punch for skin. 2–5. Tongs of various kinds with alligator jaws for mucous membrane, lymphoid tissue, fat, and muscle. 6–8. Cannula with mandrel for bone, marrow, and cartilage [42]. (Courtesy of Professor Stig Radner, Dept of Medicine, University of Lund, Sweden.)

histological examinations. Such instruments are routinely used with a great variety of organs, such as skin, mucous membranes, fat cells, lymphoid tissues, muscles, liver, spongious bone, and bone marrow.

The Radner needle is an example which illustrates the importance of having access to a proper biopsy technique [42]. This needle permits easy removal of samples of bone marrow from the posterior superior iliac crest of the pelvis. This may eliminate the need for sternal puncture, which has long been the standard technique for obtaining bone

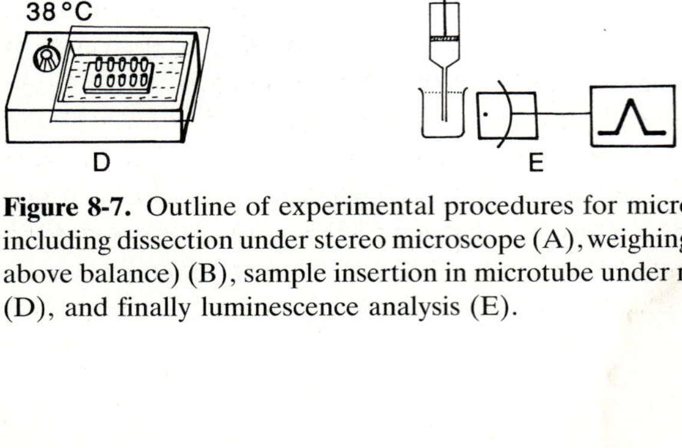

Figure 8-7. Outline of experimental procedures for microanalysis of lyophilized tissue samples, including dissection under stereo microscope (A), weighing on quartz fiber balance (scale is shown above balance) (B), sample insertion in microtube under microscope surveillance (C), incubation (D), and finally luminescence analysis (E).

marrow specimens. Sternal puncture is known to be associated with a certain, although rare, risk of causing fatal cardiac tamponade. With biopsies from the superior iliaca, bleeding is far less dangerous; except, of course, for individuals with hemophilic diseases. Since large specimens of 2 × 10 mm are obtained with the Radner needle, preparations for routine histological slides could be supplemented with or exchanged for the preparation of cryostat sections and analysis as described in the section on histochemistry (cf. Figure 8-7).

8.3.4 Cell and Organ Cultures

The limited need for material for bioluminescence analysis opens up supplementary application in investigations of cultured human cells. One such example may be the culture of fragmented fetal pancreas or isolated islets of Langerhans as transplants in certain severe diabetic conditions and where viability may be checked by luminometric methods [43]. Procedures for culturing human fat cells are also available, cf. Section 6.4.2. Recently, biopsies from gastric mucosa have been grown successfully and have been used to examine the response to hydrochloric acid secretagogues and inhibitors [44]. Parathyroid cells have been cultured from human adenomatous glands (and from bovine glands) in a study of the regulatory action of Ca^{2+} [45]. It is easy to envisage interesting applications of bioluminescence in all these cultures.

8.4 Preparative Biochemistry

In preparative biochemistry, there is a general demand for analytical processes to monitor various purification steps. There is sometimes a requirement for the analytical process to be so extremely sensitive that only small amounts of material are utilized in the monitoring process. Furthermore, there is also a requirement that the analyses should be fast to make a feedback of the analytical results of the preparative process possible. Bioluminescence analysis can fulfill these requirements quite well. With the help of coupling reactions, applications can be extended to the preparation of a great variety of substrates and enzymes. The versatility of the bacterial luciferase system in this context should be noted.

The use of bioluminescence in following the purification of various luciferases may seem to be a trivial application of bioluminescence analysis. However, the rapid analyses, which can be carried out essentially simultaneously with the various preparative steps and which require very small amounts of material, provide continuous information about the progress of the work, also yielding an essential background to collecting information about the properties of these enzymes. Thus, during the preparation of the photoprotein aequorin, the activity is conveniently monitored by the addition of Ca^{2+}. Likewise, when purifying firefly luciferase, simple addition of ATP, together with Mg^{2+} and luciferin, afford the desired information about the activity. However, with bacterial luciferase, the required substrate $FMNH_2$, which is susceptible to oxidation, must be produced during the analysis. An attractive procedure is an electrochemical setup [24]. In this case FMN is reduced at the cathode and reacts with luciferase and a long chain aldehyde to produce light, cf. Figure 8-4. This rapid analysis reduces the time required for the recognition of fractions to be collected. Dehydrogenases are contaminants in preparations of bacterial luciferase. By chromatography on a Cibachrone Blue column, it is possible to remove these enzymes. The resulting luciferase preparation may later by used to monitor purification of dehydrogenase enzymes in a single step analytical procedure [46].

8.5 Histochemistry

8.5.1 Visualization and Electronic Imaging

Histochemical staining has added biochemical information to our knowledge about the fine structure of tissues and organs. Selective staining of elements, compounds, or their functional groups is a very good way to evaluate their distribution in histological sections, which in turn is of value for studying the metabolic role of different cells. There is a large variety of applications. This may be illustrated by an example from classical histochemistry; namely, the reaction between reduced pyridine nucleotides and tetrazolium salts, which yields a blue, almost insoluble compound. Precipitation of the stain, following the reduction with pyridine nucleotides in dehydrogenase reactions, reveals sites of high activity. The recording of the native fluorescence is another way of determining the location of the reduced pyridine nucleotide [47, 48]. It has also been demonstrated that bioluminescence reactions can be used similarly. Distribution patterns of nucleotides and metabolites have been visualized in cryostat sections by coupling either to the firefly or the bacterial luciferase reactions. The blackening of a photographic film was used for imaging the reactions and for estimating their intensities [49].

In applications of bioluminescence techniques for visualization, it must be taken into account that the emission obtained may be extremely weak. Photographic registration is the conventional method of recording [50]. However, it is frequently useful to be able to directly observe the dynamics of the luminescence process. It is true that the dark adapted eye is capable of observing low levels of intensity but it should be recalled that it takes more than half an hour to reach this ability. Electronic registration and processing of the picture have great advantages and also provide ways of image intensification. It is, for instance, possible to follow intracellular changes of the calcium ion level using such means, (see Section 9.9 on probing).

8.5.2 Microanalysis

Evaluations and measurements within the visual field of a microscope can be replaced by analyses of well-defined groups of cells dissected from cryostat sections. This is largely a matter of down-scaling the analyses and developing techniques for measuring low intensities. In determinations of the composition of living cells, a crucial question is how to maintain the chemical properties during the preparation prior to analysis. It is primarily a matter of efficiently interrupting all metabolic processes. Appropriate methods were originally developed by the Carlsberg laboratory and later refined by Glick [51, 52], Lowry and coworkers [53]. Due to their work, a flexible system of microtechniques now exists with a proven ability to meet most of the demands, notwithstanding that further improvements are possible. These techniques have been successfully employed in preparations of specimens for bioluminescence analyses, but may require certain modifications. Some procedures used are discussed below.

Quick-freezing is the preferred way of stopping metabolic processes. Subsequent freeze-drying is an important step towards the success of the extraction and thus of the overall analysis. Surprisingly often, the difference between freeze-dried and fresh material is completely overlooked when different methods of extraction are compared. The freeze-dried samples may be dissected under a stereo microscope using lyophilized cryostate sections (Figure 8-7). Samples weighing a few nanograms can be dissected by hand; even in quantities of dozens, and used for bioluminescence determinations of enzymes, substrates and nucleotides. The measured products are primarily ATP and pyridine nucleotides [34-41, 54-56]. The first analyses of microdissected samples were determinations of ATP, carried out on endocrine and exocrine pancreas and on epidermis preparations. One advantage of bioluminescence measurements of pyridine nucleotide is the long experience with similar analyses, but at a different sensitivity level. These analyses are based on light absorption spectrophotometry or fluorometry. In the case of the reduced pyridine nucleotide, use of native fluorescence increases the sensitivity tenfold as compared with absorption spectrophotometry. A further tenfold gain in sensitivity is obtained by conversion of the oxidized nucleotide to highly fluorescent compounds by treatment with strong alkali. Through the use of bioluminescence, the sensitivity may be improved by several orders of magnitude. Selective destruction of one of the nucleotides is an important feature of some analytical schemes and can be carried out efficiently. Determination of the activity of an enzyme is facilitated by the inherent chemical magnification when measuring the consumption or production of the nucleotide. A similar amplification can also be accomplished for substrates and nucleotides by means of enzymatic cycles. With luminescence analysis, there is a possibility of accumulating the light signals electronically and causing effects similar to product accumulation and subsequent analysis. This is illustrated in the examples given below.

The oxidized pyridine nucleotide $NADP^+$ may be present in an extract or obtained as the measurable product in the assay of a metabolite. Let us assume that enzymatic cycles are used following the scheme given in Figure 8-8 with fluorometric or bioluminescence analysis. It can be seen that an additional analytical step is required in the fluorometric assay as compared with bioluminescence analysis. The reason is that the subsequent analysis of an accumulated measurable product is replaced by the continuous collection and storage of the pulses from the light quanta striking the photodetector. The total number of light quanta and the accumulated molecules are produced in stoichiometric amounts.

Coupled conversions between NAD^+ and NADH may be arranged for chemical amplification as exemplified in Figure 8-9. The emission from NADH in the bacterial luciferase system is here upheld by continuous reduction of formed NAD^+ in the sorbitol dehydrogenase reaction. The number of photons emitted during one cycle was

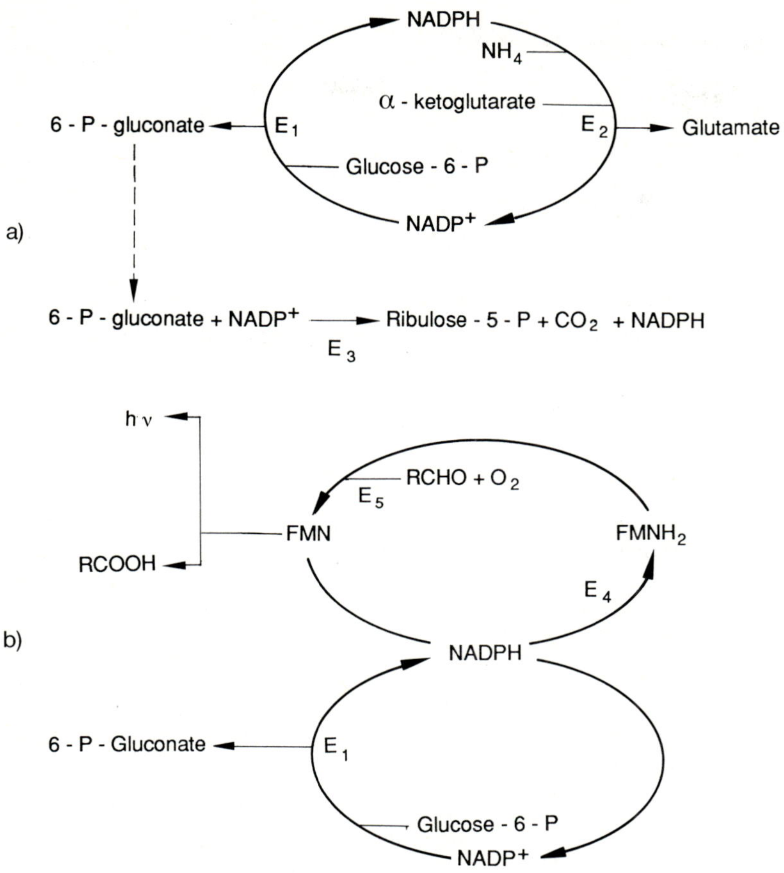

Figure 8-8. Amplification through enzymatic cycles.
a) $NADP^+$ is reduced in the glucose-6-phosphate dehydrogenase (E_1) reaction and reconverted to $NADP^+$ in the glutamate dehydrogenase (E_2) reaction. Either the oxidized or the reduced nucleotide can start the cycling and be analyzed. Amplification is obtained through product accumulation by excess substrates and letting the reaction proceed through many cycles. After stopping the cycling a final separate reaction is arranged for determining the accumulated 6-P-gluconate. Hence, 6-P-gluconate dehydrogenase (E_3) and $NADP^+$ are added, as shown below the cycle.
b) The reduction in the cycle occurs as above, but the oxidation is directly linked to the bacterial luciferase system. E_4 = oxidoreductase and E_5 = luciferase. In this system product accumulation (of 6-P-gluconate) can be replaced by photon storage (using some electronic memory device). Measures for stopping the cycling or additional reactions are not needed, as the assay is finished by setting the time for photon collection.

estimated through recordings without the presence of sorbitol dehydrogenase, and used to estimate the number of cycles as indicated on the graph [58].

In the above examples chemical amplification is caused by cyclic reconstitution of the target compound. A similar coupling is employed in order to increase the light signal in the xanthine oxidase induced luminol reaction used for labelling purposes. Here, however, the original target compound, the radical $\cdot O_2^-$ is not increased by cycling. It is

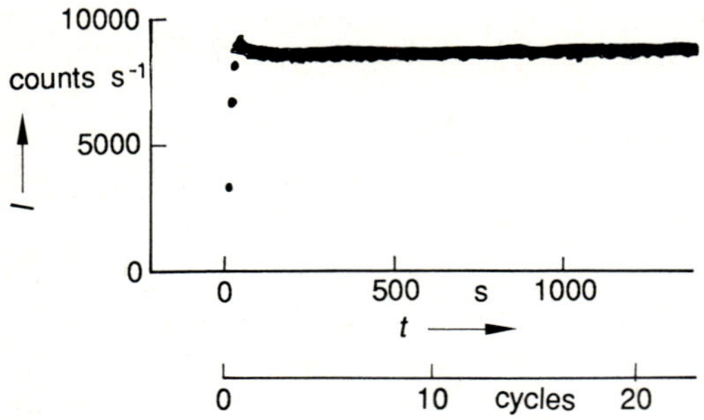

Figure 8-9. Analysis of NADH using coupled reactions arranged for chemical amplification. The system is:

$$NAD^+ + \text{Sorbitol} \xrightarrow{\text{Sorbitol dehydrogenase}} NADH$$

$$R + NADH \xrightarrow{\text{Oxidoreductase/luciferase}} \text{light}$$

Before analysis the reaction solution contains all reactants and enzymes except the pyridine nucleotide. The reaction is started by adding 5 pmole NADH. A steady-state intensity is rapidly obtained. A photon-counting equipment is used for the light recording. Each dot represents the number of photons collected during 3 s. Amplification may be obtained by collecting the total amount of light emitted (integrating the emission intensity through electronic storage of the pulses). From [58].

instead replaced by the more potent hydroxyl radical, · OH, obtained by coupling to conversions between EDTA-Fe^{3+} and EDTA-Fe^{2+} [59, 60].

References

[1] DeLuca M. In: Serio M, Pazzagli M, eds. Luminescent Assays Perspectives in Endocrinology and Clinical Chemistry. New York: Raven Press; 1982: 115–123.
[2] Wannlund J, Egghart H, DeLuca M. In: Serio M, Pazzagli M, eds. Luminescent Assays Perspectives in Endocrinology and Clinical Chemistry, New York: Raven Press; 1982: 125–128.
[3] De Wet JR, Wood KV, DeLuca M, Helinski DR, Subramani S. In: Schölmerich J, Andreesen R, Kapp A, Ernst M, Woods WG, eds. Bioluminescence and Chemiluminescence New Perspectives. Chichester: John Wiley & Sons; 1987: 369–376.
[4] De Wet JR, Wood KV, Helinski DR, DeLuca M. In: DeLuca MA, McElroy WD, eds. Methods in Enzymology, New York: Academic Press; 1986: 3–14. (Bioluminescence and Chemiluminescence, Part B; vol 133).
[5] Mulkerin MG, Wampler JE. In: DeLuca MA, ed. Methods in Enzymology, New York: Academic Press; 1978: 375–381. (Bioluminescence and Chemiluminescence; vol LVII).
[6] Wampler JE. In: DeLuca M, McElroy WD, eds. Bioluminescence and Chemiluminescence and Analytical Applications. New York: Academic Press; 1981: 249–256.
[7] Michelsen AM. In: DeLuca MA, ed. Methods in Enzymology. New York: Academic Press; 1978: 385–406. (Bioluminescence and Chemiluminescence; vol LVII).

[8] Holm-Hansen O, Karl DM. In: DeLuca MA, ed. Methods in Enzymology. New York: Academic Press; 1978: 73–85. (Bioluminescence and Chemiluminescence; vol LVII).
[9] Karl D. Microbiol Rev. 1980; 44: 739–796.
[10] Lundin A. In: Kricka LJ, Carter TJN, eds. Clinical and Biochemical Luminescence. New York: Marcel Dekker; 1982: 43–74.
[11] Van de Werf H, Verstraete W. In: Kricka LJ, Stanley PE, Thorpe GHG, Whitehead TP, eds. Analytical Applications of Bioluminescence and Chemiluminescence. New York: Academic Press; 1984: 33–48.
[12] Harber MJ. In: Kircka LJ, Stanley PE, Thorpe GHG, Whitehead TP, eds. Analytical Applications of Bioluminescence and Chemiluminescence. New York: Academic Press; 1984: 3–15.
[13] Stannard CJ, Williams AP. In: Schölmerich J, Andreesen R, Kapp A, Ernst M, Woods WG, eds. Bioluminescence and Chemiluminescence New Perspectives. Chichester: John Wiley & Sons; 1987: 481–490.
[14] Thore A, Åhnsén S, Lundin A, Bergman S. J Clin Microbiol. 1975; 1: 1–8.
[15] Åhnsén S, Lundin A, Nilsson L, Thore A. In: Schram E, Stanley PE, eds. Analytical Applications of Bioluminescence and Chemiluminescence. Westlake Village, California: State Printing & Publishing. 1979: 438–445.
[16] Chapelle EW, Piccolo GL, Deming JW. In: DeLuca MA, ed. Methods in Enzymology. New York: Academic Press; 1978: 65–72. (Bioluminescence and Chemiluminescence; vol LVII).
[17] Curtis GDW, Johnston HH, Nichols WW. In: Kricka L, Stanley PE, Thorpe, GHG, Whitehead TP, eds. Analytical Applications of Bioluminescence and Chemiluminescence. New York: Academic Press; 1984: 21–24.
[18] Hanna BA. In: DeLuca MA, McElroy WD, eds. Methods in Enzymology. New York: Academic Press, 1986: 22–27. (Bioluminescence and Chemiluminescence, Part B; vol 133).
[19] McWalter PW. In: Kricka LJ, Stanley PE, Thorpe GHG, Whitehead TP, eds. Analytical Applications of Bioluminescence and Chemiluminescence. New York: Academic Press; 1984: 17–20.
[20] Nilsson L, Åhnsén S. In: Schölmerich J, Andreesen R, Kapp A, Ernst M, Woods WG, eds. Bioluminescence and Chemiluminescence New Perspectives. Chichester: John Wiley & Sons; 1987; 491–494.
[21] Skog S, Tribukait B, Sundius G. Exp Cell Res. 1982; 141: 23–29.
[22] Lundin A, Hasenson M, Persson J, Pousette Å. In: DeLuca MA, McElroy WD, eds. Methods in Enzymology. New York: Academic Press; 1986: 27–42. (Bioluminescence and Chemiluminescence, Part B; vol 133).
[23] Stanley PE. In: DeLuca MA, McElroy WD, eds. Methods in Enzymology. New York: Academic Press; 1986: 14–22. (Bioluminescence and Chemiluminescence, Part B; vol 133).
[24] Wettermark G, Brolin SE. In: Schram E, Stanley P, eds. Analytical Applications of Bioluminescence and Chemiluminescence. Westlake Village, California: State Printing & Publishing; 1979: 212–220.
[25] Harber MJ. In: Kricka LJ, Carter TJN, eds. Clinical and Biochemical Luminescence. New York: Marcel Dekker; 1982: 189–218.
[26] Lundin A. In: DeLuca MA, ed. Methods in Enzymology. New York: Academic Press; 1978: 56–65. (Bioluminescence and Chemiluminescence; vol LVII).
[27] Lundin A, Lindberg K, Nordlander R, Nyquist O, Styrelius I. In: Schram E, Stanley P, eds. Analytical Applications of Bioluminescence and Chemiluminescence. Westlake Village, Cal: State Printing & Publishing; 1979: 467–478.
[28] Wulff K. In: Van Dyke K, ed. Bioluminescence and Chemiluminescence: Instruments and Applications, Volume II. Florida: CRC Press; 1985: 247–251.

[29] Brolin SE. In: Bergmeyer HU, ed. Methods of Enzymatic Analysis Volume III. Weinheim: Verlag Chemie; 1983: 540–559.
[30] Carrico RJ, Johnson DR, Boguslaski RC. In: DeLuca MA, ed. Methods in Enzymology. New York: Academic Press; 1978: 113–122. (Bioluminescence and Chemiluminescence, vol LVII).
[31] Reichard DW, Miller Jr RJ. In: DeLuca MA, McElroy WD, eds. Bioluminescence and Chemiluminescence Basic Chemistry and Analytical Applications. New York: Academic Press; 1981: 667–672.
[32] Wannlund J, DeLuca M. In: DeLuca MA, McElroy WD, eds. Bioluminescence and Chemiluminescence Basic Chemistry and Analytical Applications. New York: Academic Press; 1981: 693–696.
[33] Prasher DC, McCann RO, Longiaru M, Cormier M. In: Schölmerich J, Andreesen R, Kapp A, Ernst M, Woods WG, eds. Bioluminescence and Chemiluminescence New Perspectives. Chichester: John Wiley & Sons; 1987: 365–368.
[34] Hammar H. Acta Derm Venerol (Stockh). 1973; 53: 251–258.
[35] Hammar H. Arch Derm Forsch. 1973; 247: 295–308.
[36] Hammar H. J Invest Derm. 1975; 65: 315–319.
[37] Brolin SE, Wettermark G, Hammar H. Strahlenther. 1977; 153: 124–131.
[38] Wettermark G, Brolin SE, Hjertén S. Cell Mol Biol. 1977; 22: 329–339.
[39] Hammar H, Hammar L. Arch Dermatol Res. 1979; 266: 17–25.
[40] Hammar H, Alvares O. Scand J Dent Res. 1974; 82: 85–92.
[41] Hammar H, Wettermark G, Wladimiroff W. Scand J Dent Res. 83: 375–381.
[42] Dombernowsky P, Worm AM, Hainou B, Hansen HH, Nissen NJ. Scand J Haemat. 1974; 12: 270–273.
[43] Sandler S, Andersson A, Schnell A. Diabetes. 1985; 34: 1113–1119.
[44] Mård S, Norberg L, Ljungström M, Wollert S, Nyrén O, Gustavsson S. Acta Physiol Scand. 1985; 123: 349–354.
[45] Nygren P. Functional Dedifferentiation of Parathyroid Cells in Hyperparathyroidism and During Culture [Dissertation]. University of Uppsala; 1987
[46] Brolin SE. Mol Cell Biochem. 1983; 55: 177–182.
[47] Panten U, Poser W, Hasselblatt A. Diabetologia. 1970; 6: 643.
[48] Panten U, von Kriegstein E, Poser W, Schönborn J, Hasselblatt A. FEBS Lett. 1972; 20: 225–228.
[49] Paschen W. In: Schölmerich J, Andreesen R, Kapp A, Ernst M, Woods WG, eds. Bioluminescence and Chemiluminescence New Perspectives. Chichester: John Wiley & Sons; 1987: 547–550.
[50] Kricka LJ, Thorpe GHG. In: DeLuca MA, McElroy WD, eds. Methods in Enzymology. New York: Academic Press; 1986: 404–419. (Bioluminescence and Chemiluminescence; vol 133).
[51] Glick D. Quantitative Chemical Techniques of Histochemistry. New York: John Wiley and Sons; 1961.
[52] Glick D. J Histochem. 1977; 25: 1087–1101.
[53] Lowry OH, Passoneau JV. A Flexible System of Enzymatic Analysis. New York: Academic Press; 1972.
[54] Östenson CG, Ågren A, Brolin SE, Petersson B. Diabete & Metab. 1980; 6: 5–11.
[55] Östenson CG, Andersson A, Brolin SE, Petersson B, Hellerström C. In: Foa PP, Bajaj JS, Foa NL, eds. Glucagon: Its Role in Physiology and Clinical Medicine. New York: Springer Verlag; 1977: 243–254.
[56] Brolin SE. In: Serio M, Pazzagli M, eds. Luminescent Assays Perspectives in Endocrinology and Clinical Chemistry. New York: Raven Press; 1982: 63–77.
[57] Brolin SE. J Biochem Biophys Meth. 1986; 13: 179–184.

[58] Brolin SE, Naeser P, Wettermark G. In: Stanley PE, Kricka LJ, eds. Bioluminescence & Chemiluminescence Current Status. Chichester: John Wiley & Sons; 1991: 523–526.
[59] Baret A, Fert V, Aumaille J. Anal Biochem. 1990; 187: 20–26.
[60] Baret A, Nicolas JC, Perrot C. In: Stanley PE, Kricka LJ, eds. Bioluminescence & Chemiluminescence Current Status. Chichester: John Wiley & Sons; 1991: 191–194.

9 Samples of Application

9.1 Cell Cultures

Physiological and pharmacological research is nowadays frequently carried out at the cellular level. With cell cultures, there is often a desire to be able to obtain some quantitative measure of the cells. The traditional counting of the number of cells may sometimes be cumbersome and alternative methods are sought. Experiments have shown that there is a good quantitative correlation between the number of cells and the amount of ATP determined by bioluminescence methods. Such relations are discussed in somewhat more detail in Section 8.2 under the heading "Biomass Determination in Microbiology and Cell Research".

When analyzing the chemical content of cell cultures, there is usually ample material and no particular requirement for high sensitivity of the analytical methods. However, this is not always the case. The problem may, for instance, be a matter of having access to only a small number of cells when starting the culture, slow cell growth, or the disadvantage of requiring many generations for reproduction. Such a situation may arise with some parasitological and microbiological problems. The great value of microtechniques in this context was, for instance, pointed out by an international assembly arranged by the World Health Organization and Akademia Nauk [1].

Today, several mammalian cell lines are available for cell culture studies. Bioluminescence measurements have, for instance, been used in this context to analyse adenine nucleotides in Ehrlich ascites tumor cells [2] and in human prostatic carcinoma cells [3]. Studies on cultured mammalian cells are rapidly expanding and backed up by commercially available equipment and culture media.

9.2 Single Cell Assay

9.2.1 General Aspects

The importance of referring physiological and pathological processes to the cellular level was recognized early and formalized in 1856 by Virchow, who introduced the concept of the "Cellularpathologie". Analyses of the chemical content of a single cell are of value from two particular viewpoints. A cell is both a unit and a unity. The first aspect is based on the fact that cells of the same species contain and maintain the same amount of DNA, except during division. Considering this aspect, biochemical analyses may refer to the cell as a unit. The second aspect is illustrated by the fact that single cell analyses make it possible to follow the life cycle of the cells. Only in the case of cultures brought to synchronous division, somewhat similar information can be derived from mean values of analytical samples.

9.2.2 Determination of the Cell Mass

Although the cell itself is a unit, there may be a need to measure the cell mass. Scanning interference techniques do not alter the chemical composition and appear to be a useful procedure for such measurements [4, 5]. Thus, in different kind of cells ranging in weight from 0.07 to 2.5 ng, the dry mass was determined [6], see Figure 9-1.

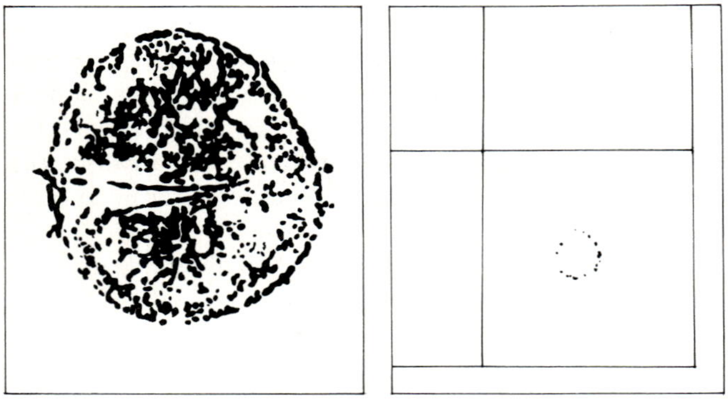

Figure 9-1. Quantification of single cells (principle). Appearance of a large cell, 2.5 ng (dinoflagellate), and a small cell, 0.07 ng (macrophage), positioned for mass determination under a scanning interference microscope.

9.2.3 Absorption Photometric and Fluorometric Measurements

When it comes to chemical analysis, some procedures now exist for analyzing certain components of single cells. They are primarily based on microphotometric techniques. Absorption photometry may be employed but requires high concentrations of compounds which possess or can be stained to yield high absorption [7, 8]. Several cytofluorometric methods have been introduced [9], and such techniques are used for the determination of enzyme activities as applied in prenatal diagnostics [10–12]. It has been demonstrated in different contexts that fluorescence analysis can satisfy varying analytical demands in single cell assay.

9.2.4 Monitoring Discharge and Intracellular Events

The flow of ATP from erythrocytes has been followed by suspending them in a luminescent reactant solution and subjecting them to individual disruption by a laser beam [13, 14]. In the respiratory burst of neutrophil granulocytes, the discharge of free oxygen radicals from a single cell was monitored by the pholasin luminescence [15]. The sensitivity of the luminol reaction was found to be insufficient for this purpose. The luminescence system of the earthworm *Diplocardia* ought to be of special value in studies of free oxygen radicals and should probably be sensitive enough, cf. Section 8.1.

Changes in the concentration of Ca^{2+} can be monitored by intracellular deposition of aequorin, cf. Section 9.9.

9.2.5 Analysis of Cellular Extracts

For an analysis at the low concentrations prevailing in an extract from a single cell, contamination of reactant solutions with bioluminescence-active components is a frequent cause of failure. Careful purification of reagents and solvents is obviously needed. In a standard laboratory, air pollution may be sufficient to ruin an analysis, even if a solution is left uncovered for only a few minutes. Filtering the air for removal of airborne organisms may be sufficient, but access to a sterile room and use of protective clothing is an advantage. An example of an analysis of the ATP content of single cells [6] is given in Figure 9-2. Special handling routines may also be developed which are less

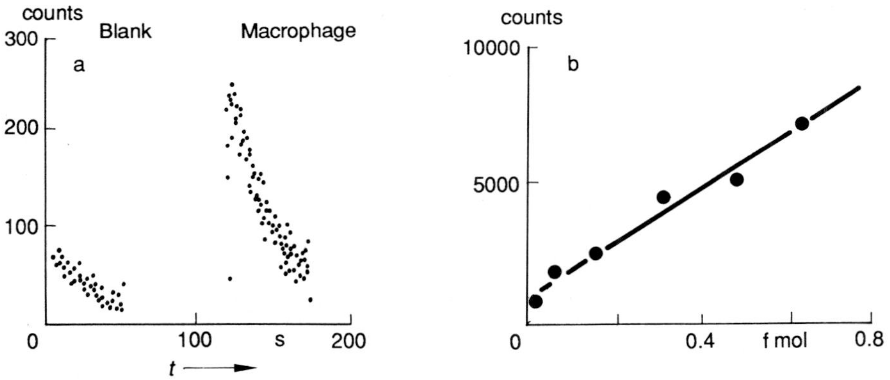

Figure 9-2. Analysis of the ATP content in an isolated mouse macrophage.
Photon counting is employed, and the cell yields about 8500 pulses during 100 channels of 0.5 s, corresponding to 0.8×10^{-15} moles. The graph to the right is the calibration curve.

susceptible to contamination. The technique using micro droplets below an oil layer, which is used in microfluorometric enzyme assay [10–12], may be a fruitful line of development. Figure 9-3 shows possible arrangements for bioluminescence analysis.

Extrapolating from the abovementioned analyses and using amplification procedures of the type mentioned in Chapter 8, it should also be possible to analyze pyridine nucleotides in single cells. Analyses of enzyme activities should also be facilitated by the inherent chemical amplification through product accumulation in enzyme reactions.

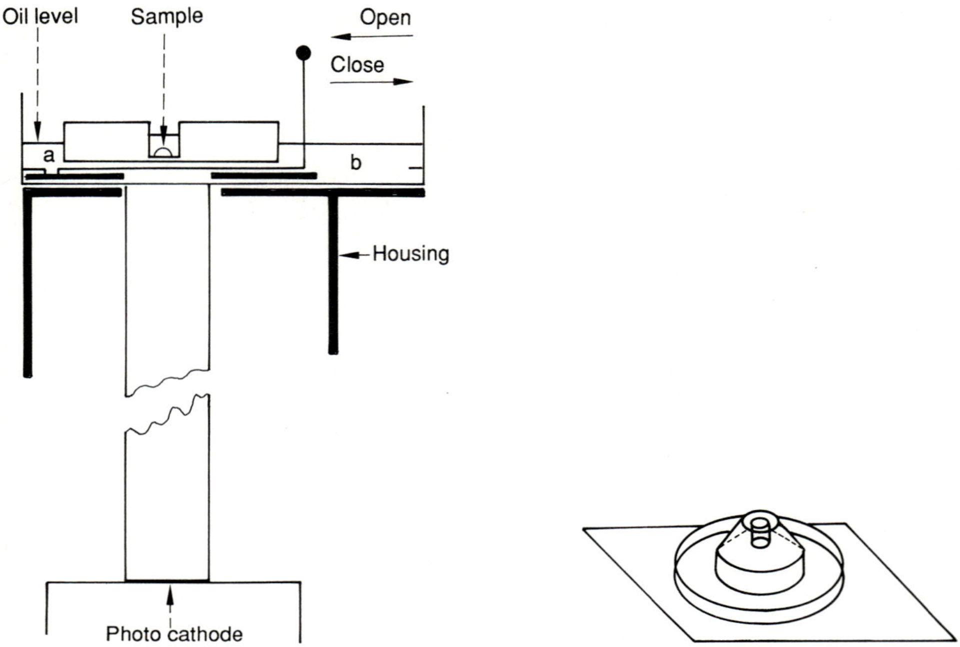

Figure 9-3. Arrangement for luminescence analysis of single cells.
Left: The sample is seen at the top as a droplet below oil in a cylindrical hole drilled in a transparent block. This is placed in a double bottomed compartment (drawn with fine lines), which likewise contains oil. A shutter is moved in the interface at the bottom through a slit (b), while a hole (a) makes way for the oil. The compartment rests on a housing so that its transparent bottom is optically connected to the light guide, leading to the cathode of a photomultiplier. The connections are arranged with a refractive index near that of the light guide. At the phototube end, the connection must allow chilling of the tube. Condensation on the light guide must be prevented. It may be lodged in a leak-proof evacuated case (not shown), or silver coated.
Right: Holder for analysis in a droplet below the oil surface, which may improve the collection of emitted light when replacing the trough shown in the left drawing. Note that lateral losses of light are minimized by preventing the oil from reaching the sides of the chamber.

9.3 Organ and Tissue Culture

9.3.1 Demands on Specimens

The metabolic state of an organ kept in a culture is dependent on its interaction with the environment. This obviously takes place through the surface of the cultured explant. Generally, a large surface area facilitates the supply of oxygen and nutrients and, likewise, the disposal of waste products. In this regard, the shape of the preparation is of great importance. Smaller specimens on the whole have a higher surface-to-mass ratio and provide a more favorable transfer of all such agents. With high demands on cellular energy production, greater dependence on an exchange with the culture medium may be expected. Accordingly, it is important to be able to analyze the contents of small specimens. With these, the high sensitivity of bioluminescence methods is of great value.

Epithelial and endothelial layers are examples of tissues which have shapes well suited for being kept alive in a culture medium. This, for instance, holds for the epidermis, the mesentary, and the innermost layer of tubular organs or body cavities. Sometimes, nature itself maintains culture medium conditions to keep certain cells viable. This is the case with the eye lens. The aqueous humor provides a medium for the lens which is nourished through the wide surface of the capsule epithelium, a situation which lends itself to mimesis in vitro and to the application of bioluminescence analysis [16].

9.3.2 Suitable Specimens, Examples and Suggestions

The islets of Langerhans have a very favorable surface-to-volume ratio, and the organ is well suited for culture experiments. In all higher species, the islets contain insulin- and glucagon-producing cells in aggregates which are distributed within the pancreas as numerous small bodies. Cultures of islets have been subjected to several kinds of experiments and may represent the best example for the use of organ culture as a way to evaluate the response of an organ to changes in environmental conditions; for instance, with respect to secretion in the presence of various nutrients and drugs. In this context, bioluminescence has provided analytical facilities for metabolic studies [17–22]. With many kinds of tissue, the principle of evaluating alterations in vivo after transfer to a culture should be a useful experimental model. Figure 9-4 is an example of this technique. The insulin-producing B-cells are destroyed by the injection of streptozo-

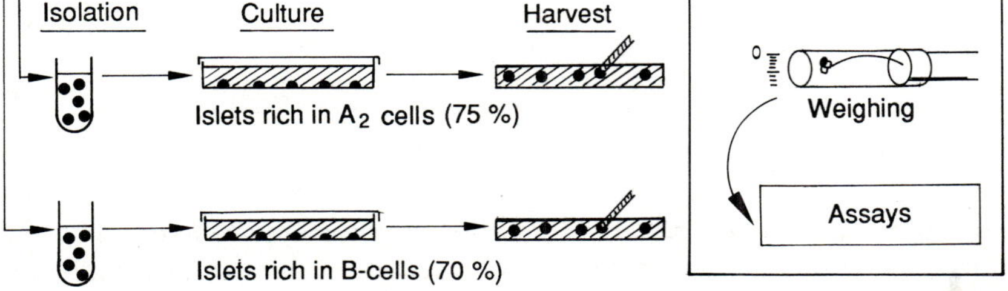

Figure 9-4. Outline of the route leading from living animal to sample suited for bioluminescence analysis via organ culture.
The example illustrates how a specific treatment in vivo can determine the conditions for a culture. In this case, streptozotocin is administered to selectively destroy the insulin-producing B cells of a guinea pig, leaving cultures rich in glucagon-producing A_2 cells.

tocin, leaving the glucagon-producing A_2-cells intact. Experience with the application of bioluminescence analysis to tissue culture has been obtained through studies of various conditions of nutritional supply in explants of human fat cells [23, 24].

One can envisage a great many situations, in which the principles described above can be used to elicit some crucial information. One example is biopsy samples of the epidermis after its exposure to radiation or the absorption of drugs. Experience with analyzing fresh biopsies of epidermis and mucous membranes is available. References are given in Section 8.3, which also contains information about other cultured specimens that are suitable for similar studies. For cell culture experiments, it may be of interest to know that a reason for the gradual decrease in the functions of the lens, in particular its accomodation ability, is related to inefficient transfer of nutrients and degradation products.

9.4 Experimental Embryology

9.4.1 Initial Stages of Development

An essential part of our knowledge on early development emanates from experiments on eggs of the sea urchin (*Enchonidia*). Here, bioluminescence provides a useful tool in evaluating the bioenergetics of the initial cellular organization and has been used for the analysis of ATP [25]. The ATP concentration has been monitored during the early development of the sea-squirt [26], i.e., the tunicate *Ciona intestinalis*, and likewise in fish and amphibian ova [25]. Furthermore, ATP has been determined in studies of the preimplantation development of the rabbit [27] and the mouse [28, 29]. A considerable part of these measurements were carried out on blastocysts flushed from the uterine tubes of the mated females. Morulae and oocytes were also collected and analyzed. Some ADP and AMP analyses have been included, but only little interest was paid to the adenylate pool. It appears that no attempts have been made to estimate the rates of adenylate conversions. Bioluminescence can be expected to meet analytical demands in current research on mammalian oocytes and blastocysts and assist in the bioenergetic evaluations of conditions and measures which are liable to cause metabolic disorder and to affect the viability.

9.4.2 The Cultured Embryo as an Experimental Specimen

Except for the initial phase of the development, luminescence analyses have only rarely been used in experimental embryology; although the potential applications appear to be versatile and improved by the development of organ culture techniques. Rat embryos provide a suitable experimental model [30], since the division into segments and an early part of the organogenesis takes place before vascular connection is established through the placenta. This means that the native nourishment during this phase of development can be closely mimicked in organ culture. Thus, bioluminescence analyses of sorbitol was carried out on explants of rat embryos, grown at a increasing glucose concentration in the culture medium [31]. The same sorbitol determination as used above with the organ culture has also been applied in vivo to embryos from diabetic rats [32]. The experimental procedure of the culture experiments is presented in Figure 9-5.

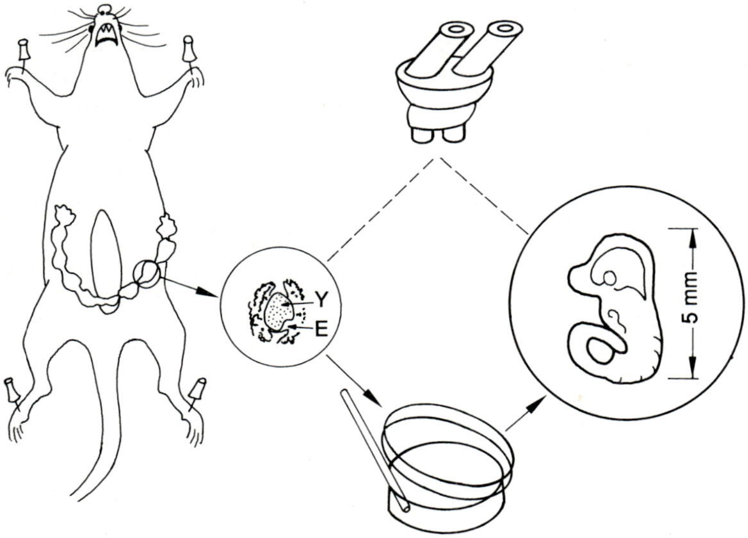

Figure 9-5. In vitro arrangement for analyses of early embryonic development.
The bicornate uterus is pulled out through an abdominal incision in an anaesthetized pregnant rat. Thickened parts are opened under the microscope, and blastocysts covered with primitive chorionic villi are peeled off. The embryo (E), attached to its yolk sac (Y), is exposed by an incision in the blastocysts and pipetted to the culture. Nourishment is supplied by the culture medium and from the yolk sac, which has been loaded with nutrients through the primitive chorionic villi. Growth in culture is allowed to proceed until segmentation and primordia of the brain, the eye, and the heart become discernible and accessible for analysis.

This model could become a convenient and inexpensive way of screening for embryotoxic effects and could to some extent replace the use of pregnant animals, which today are used in large numbers by the pharmaceutical industry for similar purposes. In a screening test with different concentrations of a drug, cultured embryos from one or two pregnant rats should suffice; whereas a similar study in vivo would require a considerable number of animals.

9.5 Laminar Elution

9.5.1 Selecting Preparative Procedures

In order to obtain the desired biochemical information, analyses are generally based on data from a fairly large number of the same kind of cells. Problems arise, if the target cells can only be collected in small numbers. In some cases, a combination of a particular preparative technique with a sensitive analytical method is applicable. The islets of Langerhans, which are dispersed as millions of corpuscles in the mammalian pancreas, may exemplify the isolation of minute small cell specimens for bioluminescence analysis. In lyophilized cryostat sections of the pancreas, samples are cut out of the islets, while the procedure is observed through a stereomicroscope, as described in Section 8.5.2. Alternatively, digestion with collagenase has been used for liberating the islets to be analyzed [17–22], a technique which has also been used with heart myocytes [33]. The cells may then be allowed to recover in a culture medium, before they are analyzed with luminescence methods. However, preparation by dissection or by digestion may fail to comply with the analytical requirements, particularly when it comes to the yield of cellular material. This is the case with thin cellular membranes, which only give access to a few cells in cross sections and are also difficult to loosen and to turn into flat preparations. This is mainly the case with cells lining tubular organs and body cavities. For the extraction of such layers, laminar elution has been developed, permitting a great variety of microanalyses [34–36].

9.5.2 Performance and Necessary Precautions

Figure 9-6 illustrates the basic principles of the laminar elution technique. After exposing the tubular organ, cannulas are inserted into both ends of the section of the organ to be assayed. Before starting the elution by a detergent acting as a membrane

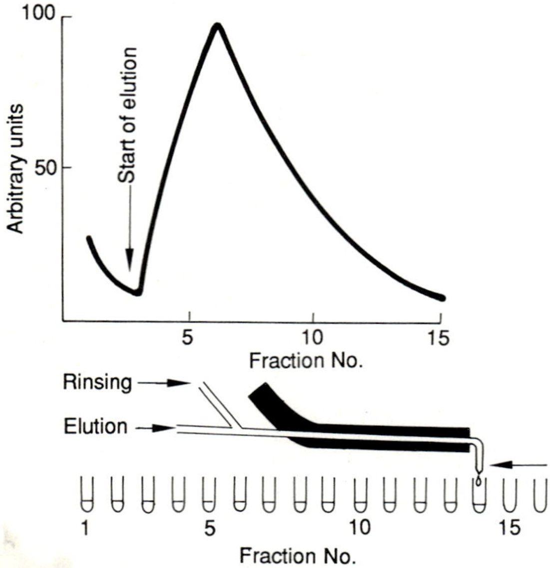

Figure 9-6. Principle of laminar elution for analysis of surface layers.
A tubular organ is cannulated, preferably in the living animal. After rinsing, the elution is started (arrow) by infusing a detergent, which acts as a membrane destroyer. Following the gradual action of the detergent on the linings of the organ exposed between the two cannules, the cell contents appear in the eluent and may be subjected to various analyses. The figure shows typical results in cases where a substance is released from the innermost layer of the lining.

destroyer, the organ is rinsed mildly with physiological saline to remove impurities. The rinsing has to be continued, until cells and cell fragments are removed sufficiently from the inner surface, as shown by the analyses of the effluent. The elution of blood vessels has to be preceded by heparinization. Furthermore erythrocytes are destroyed by rinsing with 0.85 % ammonium chloride, as in automatic blood cell counters. Their removal is conveniently checked by microanalytical hemoglobin determination [35], developed by adjusting a method based on isoluminol luminescence and designed for measuring weak hemolysis in sera [37]. The analytical principle consists in the catalysis of the isoluminol emission by Fe(II)tetrapyrrole and can also be used to recognize elution from the muscular wall of the artery by measuring myoglobin discharge.

The elution procedure is sufficiently standardized by a programmed precision pump operating at a preset speed. High reproducibility allows the eluted amounts to be well characterized in terms of the elution time and helps to identify differences in elution patterns. Reproducibility in time is obviously necessary in quantifications based on pooling samples from several elutions, unless the measurements solely concern the total elutable amount.

9.5.3 Choice of Membrane Destroyer

Elution from the interior cell layer is started by a detergent which destroys the cell membranes. SDS (sodiumdodecylsulfate, i.e., laurylsulfate) is quite sufficient for this purpose but precludes enzyme analyses, because of its action on proteins. Triton X-100 may be chosen, but not, if microspectrophotometric monitoring of the outflux is desired. Saponin is useful for the elution of nucleotides and enzymes (Figure 9-7), but its glycoside moiety seems to interfere with sugar analyses and likewise with fluorometric DNA assay, which is based on a reaction with deoxyribose. The compound G 3707 (Atlas Chemie, Everberg, Belgium [38]) does not disturb analyses of enzymes and also permits the assay of slowly degraded carbohydrates, such as sorbitol [39]. ATP, NAD^+, and NADH in the effluent are determinable, but the analyses fail to reflect the intracellular content as a consequence of the conversions within the respective nucleotide pools. It is therefore advisable to restrict the measurements to the adenine and pyridine nucleotide sums. Furthermore, it should be recalled that there is also a gradual decrease of the adenylate sum in dead tissue, as illustrated by a comparison between the measurements on two specimens of ureter, see Figure 9-7a.

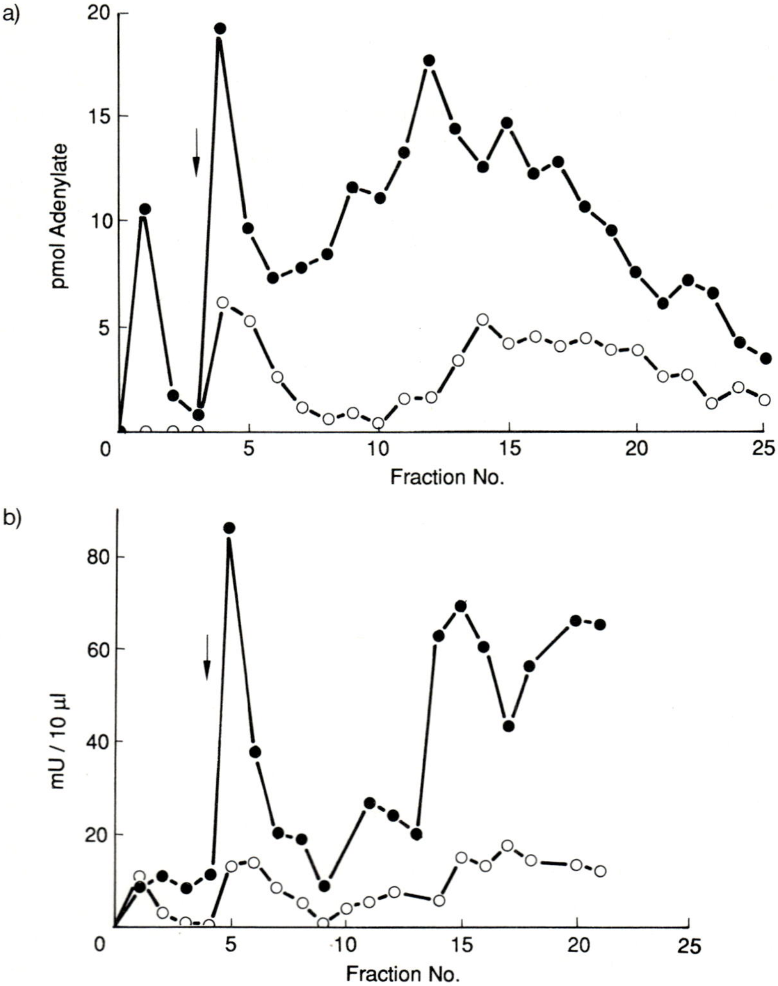

Figure 9-7. Laminar elution of rat ureter.
The first fractions represent rinsing with physiological saline, whereupon elution is started with a solution of 0.05 % saponin in deionized water at the time indicated by the arrow. At intervals of 40 s, volumes of 40 µL are infused, each yielding a new fraction.
a. Elution of adenylates (ATP + ADP + AMP) in an anaesthetized rat (filled symbols) and from the second ureter, 3 hours after killing the animal (open symbols). The two maxima are interpreted as discharge from two different epithelial layers. A marked reduction of adenylates occurs after 3 hours of ischemia.
b. Elution of enzymes, exemplified by measurements of malate dehydrogenase (filled symbols) and lactate dehydrogenase (open symbols) in an anaesthetized rat.
Again two maxima are evident.

9.5.4 Modified Applications

The idea of time-dependent laminar elution can be modified in different ways; for instance, if applied to the lining of body cavities. Modifications other than those required for surface analyses may be developed. It has been possible to produce an artificial tubular body cavity by inserting a double walled needle [40]. The needle has a closed tip and a perforated outside wall. The cells covering the holes of this wall correspond to the innermost layer of a tubular organ, yielding the same kind of initial elution profiles. Application to a pig's muscle showed that the ATP outflow was considerably lower when the test was repeated one hour after killing the animal, Figure 9-8; whereas the adenylate sum remained essentially unchanged, although accompanied by a marked change of the time course of the efflux.

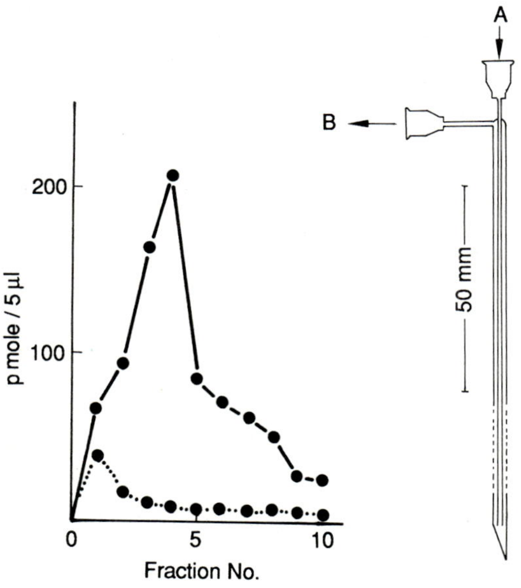

Figure 9-8. Extraction of pig muscle by perfusion with saponin.
Left: Efflux of ATP in a living pig (top curve) and in a pig one hour after death (bottom curve).
Right: Drawing of needle used for the perfusion.
The muscle cells bordering the outside perforated wall of the cannula correspond to the innermost layer of the tubular organ in laminar elution.
The large yield indicates that needles of much smaller dimensions still produce enough material for analysis. From [40].

Studies with the needle elution technique [41, 42] indicate that further developments are required in order to reduce cellular losses and damages to the tissues involved. This can be achieved by reducing the needle dimensions and decreasing elution times. It may also be useful to let a small amount of detergent act for only a limited time and after its removal complete the elution with a less agressive solution, reducing the amount of cellular damage.

9.6 Bioenergetic Evaluations

9.6.1 Metabolic Energy Provision Coupled to Light Formation

At various metabolic steps, energy transfer can be linked to light production, monitored, and studied. Energy supplied by the conversion of adenine and pyridine nucleotides can thus be evaluated. For studies of flavine nucleotides similar possibilities are within reach.

In many cases, it is desirable to arrange the analytical procedures such that the recording of the light flux follows the time course of energy production. Formation of ATP by oxidative phosphorylation has been continuously followed in mitochondrial suspensions using the firefly system [43–46]. Photophosphorylation has likewise been studied by monitoring the ATP production in preparations of chloroplasts and chromatophores [47–49]. The transformation of light to chemical energy can be exemplified by recording the ATP formation in bacterial chromatophores after their exposure to a short flash of light [48], see Figure 9-9. Within a second after the light flash the formation of ATP is almost complete.

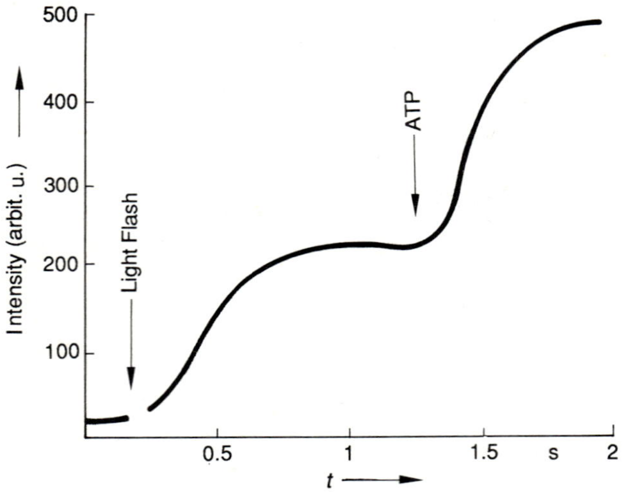

Figure 9-9. In vitro photophosphorylation studies.
Phosphorylation is induced by a flash of light directed onto chromatophors suspended in a preparation of slowly reacting firefly luciferase. ATP formation is monitored by the intensity of the resulting emission. A concentration change of 0.24 μM is obtained by injection of an internal standard, as shown on the right. The diagram is based on data from [48].

9.6.2 Evaluations Through Arrested Energy Flux

A useful procedure in bioenergetic research is the instantaneous interruption of the oxygen supply. A suitable experimental arrangement is illustrated in Figure 9-10. The steady-state ATP concentration becomes lowered and the decrease is a measure of the

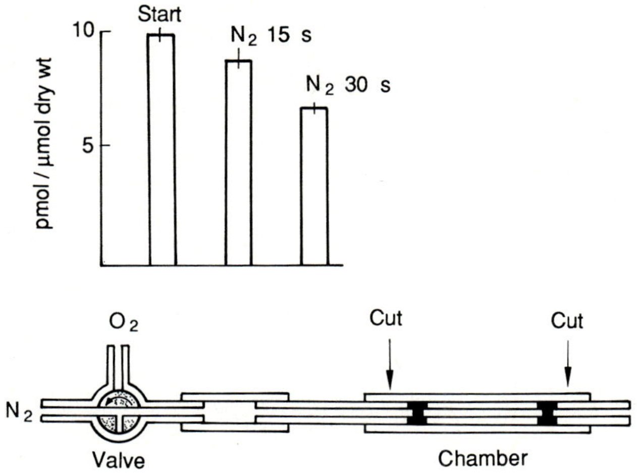

Figure 9-10. Failure of ATP production in interrupted oxidative phosphorylation.
The diagram illustrates the effect on the ATP concentration by interrupting the oxygen supply to insulin-producing cultured islets of Langerhans.
A drawing of the experimental device is shown below. Rapid exchange from oxygenated to deoxygenated perfusion medium is achieved by simply turning the stop-cock. The perfusion chamber is made by pressing polyethene tubing against grids inserted in another slightly larger tubing. The experiment is terminated by quickly cutting the chamber and expelling the samples in chilled isopentane. Modiefied from [22].

efficiency of the respiration. Complementary information may be obtained by determining the oxygen consumption, for instance with Cartesian micro divers [50], as in measurements on islets of Langerhans [51], freshly prepared or kept in a culture [17]. Experimental options are also available for further elucidation of other pathways towards ATP production. Measurement of lactate accumulation is the classical method of evaluating anaerobic glycolysis. Lactate analyses calls for precautions which are also needed in bioluminescence determinations [52]. The compound antimycin arrests oxidative metabolism selectively and has been used in islet research, for studies of respiration in vitro [53].

Bioluminescence analysis is suitable for elucidating variations in the concentration of different adenylates caused by changes in the oxygen supply to the islets of Langerhans. In one set of experiments, islet samples were dissected from lyophilized sections of quickly frozen pancreas with interrupted vascular supply [54], see Figure 8-2a. In another set, isolated islets in a culture were made hypoxic by exchange to a deoxygenized medium [22] (Figure 8-2b). In all cases, there was a rapid decrease in ATP concentration; whereas the sum of the various adenylates remained unchanged. If the amount of each adenine nucleotide is determined (Figure 8-2a), the adenylate charge defined by Atkinson [55, 56], i.e., the ratio ([ATP] + 0.5 [ADP])/([ATP] + [ADP] + [AMP]) can be evaluated.

Changes of the NAD^+- and the NADH-level are also conveniently determined through luminescence. Thus, the rise in NADH concentration was studied in islet preparations from quickly frozen pancreas [57] (Figure 9-11). A redox state as defined by the ratio $[NAD^+]/[NADH]$ may also be evaluated and is a useful parameter in

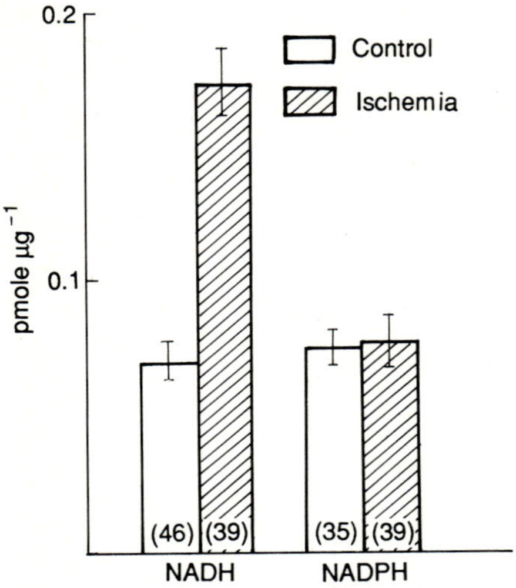

Figure 9-11. Example of the effect of ischemia on the concentration of reduced pyridine nucleotides.
Islets of Langerhans from mice, isolated from cryostat sections of pieces of pancreas frozen either freshly or after ischemia (cf. Figure 8-7). Each of the reduced nucleotide forms was assayed after selective enzymatic oxidation of the other. The higher reactivity of NADH as compared to NADPH is obvious. From [57].

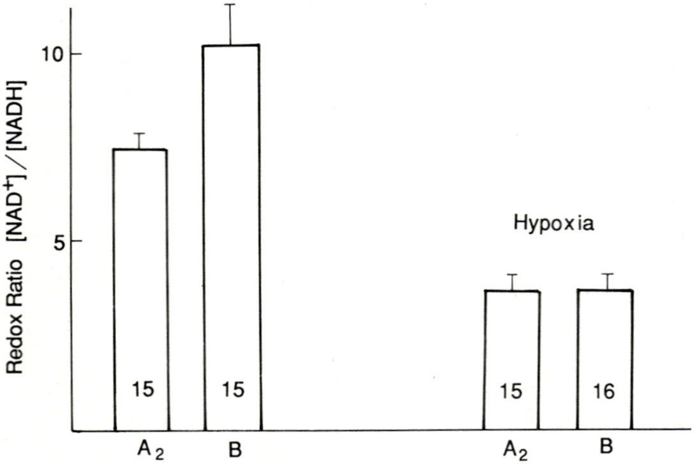

Figure 9-12. The redox state of A_2- and B-cell-rich islets of Langerhans from guinea pig.
Right: Analysis of freshly prepared specimens.
Left: Analysis of islets recovered from culture medium.
The effects of the initial hypoxia during the preparation and later recovery are clearly evident. From [58].

bioenergetic studies. One application may be in checking the condition of cultured organs, as exemplified in Figure 9-12 with freshly prepared islets. The ratio increases if the preparations are allowed to recover in a culture medium [58].

9.6.3 Inherent Weak Light Emission

All living species produce some luminescence by a mechanism that is different from that of the bioluminescence reactions used in current analytical schemes. Low intensity luminescence may, however, be used to gain information about various biological processes. It is, for instance, used to evaluate the response of species to environmental stress and to various specific factors, such as antioxidants, nutrients, etc. [59].

9.7 Metabolic Transfer of the Sulfate Group

9.7.1 Metabolic Detoxification by Sulfatation

The formation of sulfates and glucuronates is an important part of the detoxification process. The ability to transfer the sulfate group correlates with the metabolism of the phosphoadenosine nucleotides APS, PAP, and PAPS, see Table 9-1. As indicated in reaction (R 9-3), sulfate can be transferred from PAPS to many different compounds.

Table 9-1. Conversions of adenine phosphosulfates

$SO_4^{2-} + ATP \xrightarrow{E_1} APS + PP_i$	(R 9-1)
$APS + ATP \xrightarrow{E_2} PAPS + ADP$	(R 9-2)
$PAPS + R\text{-}OH \xrightarrow{E_3} R\text{-}SO_4 + PAP$	(R 9-3)
$PAP + \text{Luciferyl sulfate} \xrightarrow{E_4} \text{Luciferin} + PAPS$	(R 9-4)
$\text{Luciferin} + O_2 \xrightarrow{E_5} \text{Oxyluciferin*} + CO_2$	(R 9-5)
$\text{Oxyluciferin*} + GFP \rightarrow \text{Oxyluciferin} + GFP*$	(R 9-6)
$GFP* \rightarrow GFP + h\nu$	(R 9-7)

APS Adenosine-5′-phosphosulfate
PAP 3′5′-phosphoadenosine phosphate
PAPS 3′-phosphoadenosine-5-phosphosulfate
E_1 ATP-sulfurylase (ATP sulfateadenyltransferase, EC 2.7.7.4)
E_2 APS-kinase (EC 2.7.1.25)
E_3 Sulfotransferase (sulfotransferases transfers sulfate to many different compounds such as steroids, phenolic compounds, aromatic amines and cerebrosides)
E_4 Luciferin sulfokinase
E_5 Luciferase (EC 1.13.12.5)
GFP Green-fluorescent protein
PAP is measured by the formation of *Renilla* luciferin in sulfate transfer from luciferyl sulfate. In the analysis of PAPS, it is first hydrolyzed to PAP.

This, for instance, is the case with steroids, phenol derivatives, aromatic amines, some other cyclic and heterocyclic compounds, and with cerebrosides. All of the three phosphoadenosine nucleotides can be brought to produce a light-inducing compound and are thus determinable through bioluminescence assay.

9.7.2 Analyses Based on the Renilla System

Addition of luciferyl sulfate and luciferin sulfokinase from the sea pansy *Renilla reniformis* results in the formation of PAPS and luciferin, as shown in reaction (R 9-4). The amount of luciferin formed is then measured by the light emission in reactions (R 9-5) – (R 9-7), in which luciferyl sulfate does not participate [60]. A convenient and sensitive assay of PAP has been developed along this line and has also been used for the determination of PAPS after a preceding hydrolysis [61, 62]. From a mechanistic point of view it is interesting that the excited oxyluciferin is not an emitter. Instead, its energy is efficiently transferred to a green-fluorescing protein, GFP.

9.7.3 Analysis Based on the Firefly System

Reaction (R 9-1) can be made to proceed in the opposite direction of that indicated in Table 9-1, yielding ATP. ATP is conveniently measured with the firefly system. In this way, the assay of ATP-sulfurylase has been possible and was applied in the determination of its activity in spinach leaves [63]. The same reaction, catalyzed by spinach ATP-sulfurylase, has also proved suitable for the assay of APS [63].

9.7.4 Aspects of Future Research

The above-mentioned analytical possibilities are not only useful when evaluating the defense against poisons and various degradation products of drugs, but may also be useful in other kinds of metabolic research. Thus, sulfate transfer from PAPS is of fundamental physiological importance and has some bearing on the field of chemical pathology, for instance with regard to the requirements for forming of neural myelin sheaths [61]. Synthesis of appropriate *Renilla* luciferylsulfate has been carried out [60, 61], but assays of PAP and PAPS have been hampered by difficult access to the *Renilla* enzymes. Future research on sulfate metabolism seems to be dependent on the production of the desired enzymes by gene technology or by extensive coupling to the set of reactions in Table 9-1 in such a way that ATP is obtained as a determinable product.

9.8 Discharge from Cells

9.8.1 Origin and Elimination of Extracellular Material

Determination of the intracellular content of various cells often requires the removal of extracellular material. In general, microbial cells have comparatively resistant membranes; and discharged products or necrotic debris can be removed or degraded into analytically inactive material without causing leakage through their cellular membranes. With human or other mammalian cells, extracellular material is more difficult to eliminate, because of the induced leakage of material through the less resistant membranes. In analyses of cultured cells, gentle washing is usually applied in order to eliminate extracellular matter, but may in some cases only lead to an exchange, resulting in an altered composition of the extracellular environment. When preparing pellets of cells for analysis, a possible contribution of the supernatant to the analytical response should be kept in mind.

The release of various substances may be more or less physiological. It may be the result of malconditions of various sorts, e.g., malnutrition or a lack of oxygen, but it may also be a functional physiological response to a specific stimulus. The extracellular concentrations that are usually low as compared with the intracellular content cause some difficulty with this kind of analysis. Furthermore, the amount leaking from damaged cells, though only a small part of the entire cellular content, may be large compared with the specific release. Efforts may therefore be required to minimize such a leakage in order to distinguish a true physiological release from an unspecific loss. With cultured cells, arrangements in layers instead of suspensions can diminish damage during washing. Further improvements may be possible by using special feeder layers [64].

9.8.2 Monitoring the Discharge

In the analysis of cell suspensions, extracellular matter is generally regarded a contaminant to be removed by various methods. However, the analysis of such matter may be a source of important information and, for example, help to determine the efficiency of a washing process or other means of cleaning. Analyses may also be used to quantify the extent of experimentally induced discharge from the cells. Procedures may thus be developed to record the discharge following various physical actions (Figure 9-13) and chemical effects (Figures 9-14 to 9-16) and may be used to measure the cell content or to evaluate the membrane resistance or permeability.

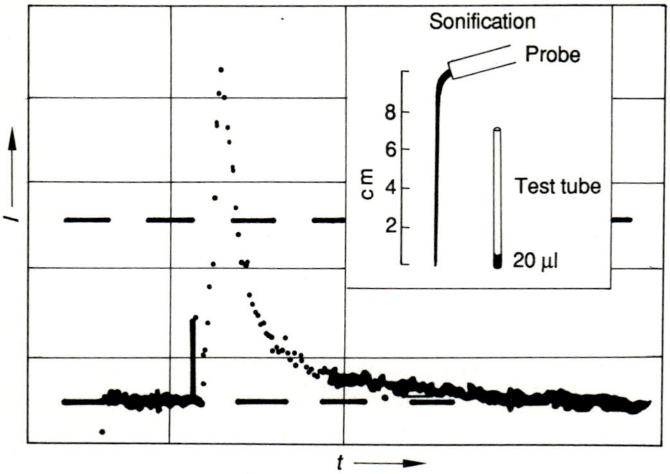

Figure 9-13. Analysis of cell content, following mechanical destruction.
A liver biopsy in 20 µL firefly luciferase solution has been subjected to sonification, which results in a rapid release of ATP, as shown by the immediate appearance of the luminescence peak 4 s after starting the sonification process. The insertion shows the probe which is connected to an ultrasonic generator ordinarily used by dentists for the removal of tartar.

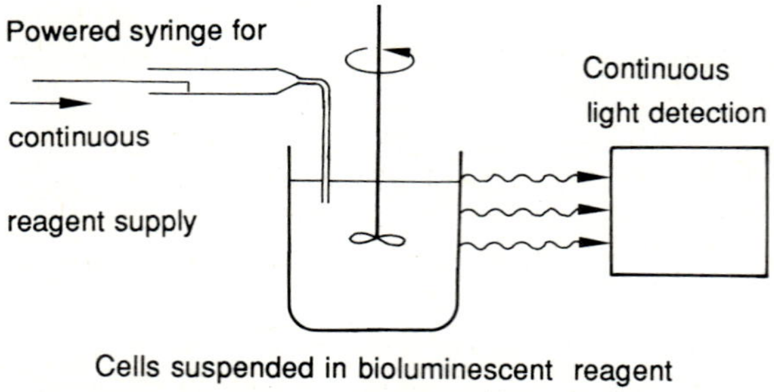

Figure 9-14. Principle of analyzing the discharge from cells, following the addition of increasing amounts of a releasing agent.

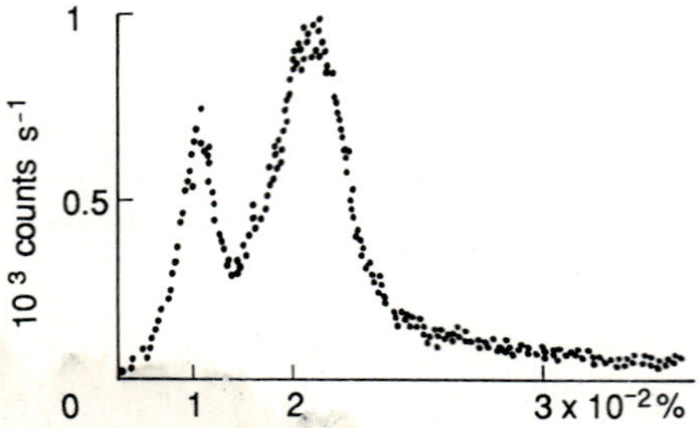

Figure 9-15. Appearance of detergent-induced cellular discharge.
ATP released when destroying cells from a buffy coat (top layer of a blood cell sediment) by addition of saponin. Ordinate: luminescence signal; Abscissa: saponin concentration. The first peak represents discharge from erythrocytes; and the second peak relates to the leucocytes. From [75].

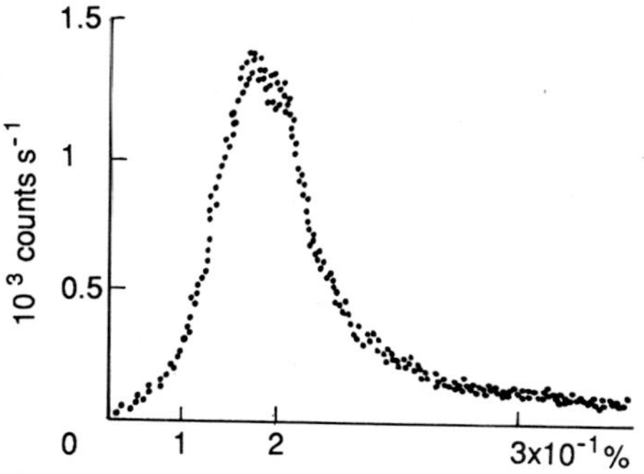

Figure 9-16. ATP release from cells of a buffy coat adding increasing amounts of the histamine liberator 48/80, which causes no detectable decrease of the leucocyte count. From [75].

9.8.3 Determination of the Physiological Discharge

The sensitivity of bioluminescence makes it possible to reveal and evaluate a small physiological discharge into its surrounding medium. Various applications can be suggested. After triggering membrane receptors, a physiological release of hormones or signal substances results. In this respect bioluminescence immunoassay may become important, see Section 8.3.2. A true release should not be accompanied by leakage of irrelevant substances. Usually cytoplasmic enzymes, which are not normally released, may be used to indicate harmful leakage. There are instances, however, when the release of an enzyme is a specific response, e.g., the release of protease induced by the action of interleukin. In this case, bioluminescence analyses should be particularly adaptable, as bacterial luciferase is highly susceptible to degradation by protease [65–67].

In studies of the respiratory burst of granulocytes and other phagocytic cells, the discharge of oxygen radicals is generally monitored by the light emission from oxidized luminol (cf. Section 1.4). The sensitivity of bioluminescence can be considerably increased by exchanging luminol for pholasin [15]. Furthermore, the facilities and advantages of the *Diplocardia* luciferin-luciferase system deserve attention, as pointed out in Section 9.2.4.

9.9 Bioluminescence Probing

9.9.1 Cellular Components Employed as Labels

The use of bioluminescence for recognition and identification purposes concerns whole cells, interior components, or discharge from the cells. The question of whether cells are present in body fluids can be answered by screening for intracellular material, which thus serves as a natural marker. The content of cells in tissue samples may likewise be estimated. Thus as described in Section 9.5.2 regarding laminar elution, a marker is required to indicate when the efflux collection from the cells lining the inner surface of the tubular organ is completed, and when the extraction from the subjacent layer begins. Analyses of adenylates have been used in this connection (Figure 9-17). Their presence in the effluent indicates extraction from metabolic active layers composed of epithelial or endothelial cells.

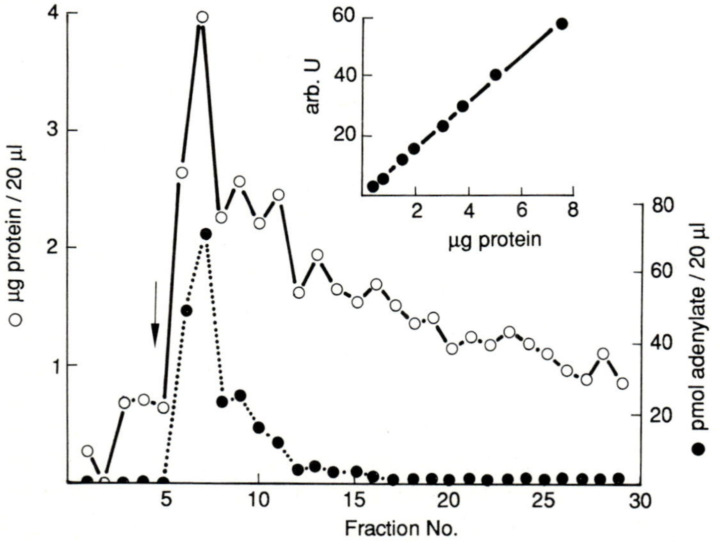

Figure 9-17. Rat aorta subjected to laminar elution with a detergent [36]. The arrow marks the start of the elution. There is an outflow of protein and adenylates (AMP + ADP + ATP). Obviously, the removal of successive layers of the aorta endothelium is accompanied by a fairly continuous protein release. The release of adenylates marks the destruction of the endothelial cell layer. The insertion is a calibration curve for protein determination with fluoresceamine. From [34].

The problems of correlating function with intracellular compartments are well-known from early studies based upon ultracentrifugation and electron microscopy. With the separation of homogenates into fractions by increasing centrifugal forces, enzymes were found to be suitable markers of intracellular sites. If, during the preparation of small samples, a similar structural recognition is wanted, bioluminescence enzyme analysis could be useful.

9.9.2 Intracellular Deposition of Markers

The technique of depositing a bioluminescent label intracellularly was first used during introduction of aequorin into a single muscle fiber [68]. A rise in intracellular concentration of Ca^{2+} was accompanied by an increased emission intensity. The method is useful for studies of muscle physiology at the cellular level [33]. Similar experiments have been carried out in a large variety of cells, such as different kinds of muscle cells, squid axons, and oocytes from various species. A comprehensive review of 130 references up to 1985 has been published [69]. The possibility of applying microscopic image intensification for recording the spacial distribution while following the changes in Ca^{2+} concentration is discussed. So far, the smallest cells yielding readily detectable signals seem to be isolated human fibroblasts [70].

Aequorin from jelly fish may be replaced by a similar photoprotein from the hydroid *Lobelia;* albeit there is a moderate loss in sensitivity. The natural aequorin sources are thought to be insufficient for large-scale applications. However, products from recombinant DNA technology may soon satisfy demands [71]. The application of gene technology to encode the production of luciferase or photoprotein is facilitated by ready control over the experiments by way of the light emission in case of success. Hence, the desired product itself serves as a probe.

In our laboratory, some attention has been paid to marking cellular organelles. A probe was introduced into mitochondria by coupling to carnitine that serves physiologically as a carrier of fatty acids across the mitochondrial membranes. In this case, linkage to isoluminol was used [72]. Lysosomes were labelled by phagocytosis of either isoluminol or luciferinylglucuronate covalently bound to Sepharose granules [73].

In a novel, theoretically interesting probing technique with possible future use, bacteriophages have been made to introduce lux genes into host bacteria [74]. The lux genes from luminescent marine bacteria were cloned and transduced to phages. These attack target bacteria within a narrow host range and introduce their DNA, including the lux genes, which encode the protein synthesis for the formation of luciferase. A particular advantage is the fact that amplification of emission intensity is achieved in two ways. First, a large number of luciferase protein molecules is obtained by the repeated utilization of the new nucleic acid template for the synthesis of macromolecules. Second, the increasing number of infected bacteria increases the production of luciferase. It is possible to identify bacterial species by light emission, which is induced by selected types of phages carrying the luminescence genes. Antibiotics, which block DNA replication or protein synthesis conducted by DNA, obviate the production of light. If phages are available which are capable of transferring the lux operon into a given type of bacteria, their susceptibility to antibiotics can be evaluated by measuring the reduction of the light emission. The introduction of lux genes by bacteriophages may illustrate the potential versatility of bioluminescence probing techniques. At the last luminescence conference, bioluminescence and imaging including expression of luciferase was an important session and forms a separate chapter of the congress volume [76].

References

[1] Sherman IW. In: Michal F, ed. Application of Biochemical Micromethods for the Investigation of Tropical Disease Pathogens. Geneva: UNDP/WORLDBANK/WHO; 1982: 37–38.

[2] Skog S, Tribukait B, Sundius G. Exp Cell Res. 1982; 141: 23–29.

[3] Lundin A, Hasenson M, Persson J, Pousette Å. In: DeLuca MA, McElroy WD, eds. Methods in Enzymology New York: Academic Press; 1986: 27–42. (Bioluminescence and Chemiluminescence, Part B; vol 133).

[4] Caspersson T, Carlsson L, Svensson G. Exp Cell Res. 1954; 7: 601–602.

[5] Petersson B. Histochem J. 1968; 1: 55–58.

[6] Wettermark G, Stymne H, Brolin SE, Petersson B. Anal Biochem. 1975; 63: 293–307.
[7] Zimmer HG. In: Neuhoff V. ed. Micromethods in Molecular Biology. Berlin: Springer Verlag; 1973: 297–328.
[8] Zimmer HG. In: Michal F, ed. Application of Biochemical Micromethods for the Investigation of Tropical Disease Pathogens. Geneva: UNDP/WORLDBANK/WHO; 1982: 133–137.
[9] Ruch F, Leeman U. In: Neuhoff V, ed. Micromethods in Molecular Biology, Berlin: Springer Verlag; 1973: 329–346.
[10] Jongkind JF. In: Michal F, ed. Application of Biochemical Micromethods for the Investigation of Tropical Disease Pathogens. Geneva: UNDP/WORLDBANK/WHO; 1982: 124–130.
[11] Jongkind JF, Hoogeveen AT. In: Michal F, ed. Application of Biochemical Micromethods for the Investigation of Tropical Disease Pathogens. Geneva: UNDP/WORLDBANK/WHO; 1982: 131–132.
[12] Galjaard H. Genetic Metabolic Diseases Early Diagnosis and Prenatal Analysis. Amsterdam: Elsevier; 1980.
[13] Weed JR. C R Acad Sci (Paris), Serie D. 1974; 278: 545–547.
[14] Lajeunie R. Nouv Rev Francaise d'Hematol Tome. 1974; 14: 774–782.
[15] Roberts PA, Campbell AK. In: Schölmerich J, Andreesen R, Kapp A, Woods WG, eds. Bioluminescence and Chemiluminescence New Perspectives. Chichester: John Wiley & Sons; 1987: 133–136.
[16] Cheng H-M, von Salza I, Gonzales RG, Ansgari NH, Srivastiva SR. Exp Eye Res. 1984: 355–364.
[17] Hellerström C, Brolin SE. In: Hasselblatt A, von Brichhausen F, eds. Handbook of Experimental Pharmacology. Berlin: Springer Verlag; 1975: 57–78.
[18] Kilbert Jr IH, Schiff MS, Foa PP. Horm Metab Res. 1972; 4: 242–244.
[19] Hedeskov CJ, Capito K. Biochem J. 1974; 140: 423–433.
[20] Östensson CG, Ågren A, Brolin SE, Petersson B. Diabete & Metab. 1980; 6: 5–11.
[21] Brolin SE, Ågren A, Petersson B. Acta Endocr. 1981; 96: 93–99.
[22] Wersäll P, Brolin SE, Petersson B, Östenson CG. In: DeLuca MA, McElroy WD, eds. Bioluminescence and Chemiluminescence Basic Chemistry and Analytical Applications. New York: Academic Press; 1981: 681–685.
[23] Kather H, Schröder F, Simon B. In: Serio M, Pazzagli M, eds. Luminescent Assays Perspectives in Endocrinology and Clinical Chemistry. New York: Raven Press; 1982: 53–56.
[24] Kather H, Wieland E, Scheurer A, Vogel G, Widenberg U, Joost C. J Clin Invest. 1987; 80: 566–572.
[25] Zotin AI, Faustzov VS, Radzinskaja LI, Ozernyuk ND. J Embryol Exp Morph. 1967; 18: 1–12.
[26] D'Anna T. Experientia. 1969; 25: 542–543.
[27] Brooks DE, Lutwak-Mann C. Nature. 1971; 229: 202–203.
[28] Quinn P, Wales RG. J Reprod Fert. 1971; 25: 133–135.
[29] Spielmann H, Jacob Muller U, Schultz P. Anal Biochem. 1981; 113: 172–178.
[30] New DAT. Biol Rev. 1978; 53: 81–122.
[31] Eriksson UJ, Brolin SE, Naeser P. Diabetes Res. 1989; 11: 27–32.
[32] Eriksson UJ, Naeser P, Brolin SE. Diabetes. 1986; 35: 1356–1363.
[33] Cobbold PH, Bourne PK. Nature. 1984; 312: 444–446.
[34] Brolin SE. J Biochem Biophys Meth. 1986; 13: 179–184.
[35] Brolin SE, Naeser P. Mol Cell Biochem. 1987; 73: 55–59.
[36] Brolin SE, Naeser P, Hjertén S. Prep. Biochem. 1987; 17: 327–336.
[37] Olsson T, Bergström K, Thore A. Clin Chem Acta. 1982; 122: 125–133.

[38] Grant DAW, Hjertén S. Biochem J. 1977; 164: 465–468.
[39] Brolin SE, Naeser P. Diabetes Res. 1988; 8: 59–61.
[40] Brolin SE, Wettermark G. In: DeLuca MA, McElroy WD, eds. Bioluminescence and Chemiluminescence Basic Chemistry and Analytical Applications. New York: Academic Press 1981: 287–293.
[41] Wulff HC, Wettermark G, Husum B. Biomed Biochim Acta 1984; 43: 1173–1177.
[42] Waaben J, Wulff HC, Wettermark G, Andersen K, Husum B. Biomed Biochim Acta. 1985: 1113–1118.
[43] Strehler BL, Totter JR. Arch Biochem. 1952; 40: 28–41.
[44] Leamaster JJ, Hackenbrock CR. Biochem Biophys Res Commun. 1973; 55: 1262–1270.
[45] Leamaster JJ, Hackenbrock CR. Eur J Biochem. 1976; 67: 1–10.
[46] Leamaster JJ, Hackenbrock CR. In: DeLuca MA, ed. Methods in Enzymology. New York: Academic Press; 1978: 36–50. (Bioluminescence and Chemiluminescence; vol LVII).
[47] Strehler BL, Hendley DD. In: Mc Elroy WD, Glass HB, eds. A Symposium on Light and Life. Baltimore: Johns Hopkins Press; 1961: 601–608.
[48] Lundin A, Thore A, Baltscheffsky M. FEBS Lett. 1977; 79: 73–76.
[49] Lundin A, Baltscheffsky M. In: DeLuca MA, ed. Methods in Enzymology. New York: Academic Press; 1978: 50–56. (Bioluminescence and Chemiluminescence; vol LVII).
[50] Holter H, Linderström-Lang K. On the Cartesian Diver, C.R. Carlsberg Ser. Chim. 24; Copenhagen; 1943: 333–478.
[51] Hellerström C. Biochem J. 1966; 98: 7C–9C.
[52] Robrish SA, Curtis MA, Sharer SA, Bowen WH. Anal Biochem. 1984; 136: 503–508.
[53] Edwards JC, Hellerström C, Petersson B, Taylor KW. Diabetologia. 1972; 8: 93–98.
[54] Brolin SE. In: Serio M, Pazzagli M, eds. Luminescent Assays Perspectives in Endocrinology and Clinical Chemistry. New York: Raven Press; 1982: 63–77.
[55] Atkinson DE. Ann Rev Microbiol. 1969; 23: 47–68.
[56] Chapman AG, Fall L, Atkinson DE. J Bacteriol. 1971; 108: 1072–1086.
[57] Berne C, Brolin SE, Ågren A. Horm Metab Res. 1973; 5: 141–142.
[58] Brolin SE, Ågren A, Petersson B. Acta Endocr. 1981; 96: 93–99.
[59] Slawinska D, Slawinski J. In: Burr JG, ed. Chemiluminescence and Bioluminescence. New York: Marcel Dekker; 1985: 533–601.
[60] Cormier MJ. In: DeLuca MA, ed. Methods in Enzymology. New York: Academic Press; 1978: 237–244. (Bioluminescence and Chemiluminescence; vol LVII).
[61] Anderson JM, Hori K, Cormier MJ. In: DeLuca MA, ed. Methods in Enzymology. New York: Academic Press; 1978: 244–257. (Bioluminescence and Chemiluminescence; vol LVII).
[62] Stanley PE, Kelley BC, Tuovinen OH, Nicholas BJD. Anal Biochem. 1975; 67: 540–551.
[63] Balharry GJE, Nicholas DJD. Anal Biochem. 1971; 40: 1–17.
[64] Freshney RI. Culture of Mammal Cells a Manual of Basic Technique. New York: Alan R. Liss Inc. 1983: 131–133.
[65] Njus D, Baldwin TO, Hastings JW. Anal Biochem. 1974; 61: 280–287.
[66] Baldwin TO. In: DeLuca MA, ed. Methods in Enzymology. New York: Academic Press; 1978: 198–201. (Bioluminescence and Chemiluminescence; vol LVII).
[67] Ziegler MM, Rausch SK, Merritt MV, Baldwin TO. In: Schölmerich J, Andreesen R, Kapp A, Ernst M, Woods WG, eds. Bioluminescence and Chemiluminescence New Perspectives. Chichester: John Wiley & Sons; 1987: 377–380.
[68] Ridgway EG, Ashley CC. Biochem Biophys Res Commun 1967; 29: 229–234.
[69] Blinks JR. In: Vandyke K ed. Bioluminescence and Chemiluminescence: Instruments and Applications; Volume II. Florida: CRC Press; 1985: 185–226.
[70] Cobbold PH, Goyns H. Biosci Rep. 1983; 3: 79–86.
[71] Prascher DC, McCann RO, Longiaru M, Cormier M. In: Schölmerich J, Andreesen R, Kapp A, Ernst M, Woods WG eds. Bioluminescence and Chemiluminescence New Perspectives. Chichester: John Wiley & Sons; 1987: 365–368.

[72] Lippman RD. In: Serio M, Pazzagli M, eds. Luminescent Assays Perspectives in Endocrinology and Clinical Chemistry. New York: Raven Press; 1982: 215–219.
[73] Lippman RD, Ågren A, Uhlen M. Mech Ageing Dev. 1981; 17: 283–287.
[74] Ulitzur S, Kuhn J. In: Schölmerich J, Andreesen R, Kapp A, Ernst M, Woods WG, eds. Bioluminescence and Chemiluminescence New Perspectives. Chichester: John Wiley & Sons; 1987: 463–472.
[75] Wettermark G, Brolin SE, Juhlin L. J Colloid Interface Sci. 1980; 73: 287–289.
[76] Session IV Bioluminescence and Imaging. In: Stanley PE, Kricka LJ, eds. Bioluminescence & Chemiluminescence, Current Status. Chichester: John Wiley & Sons; 1991: 335–368.

Index